Springer Proceedings in Mathematics & Statistics

Volume 126

More information about this series at http://www.springer.com/series/10533

Springer Proceedings in Mathematics & Statistics

This book series features volumes composed of select contributions from workshops and conferences in all areas of current research in mathematics and statistics, including operation research and optimization. In addition to an overall evaluation of the interest, scientific quality, and timeliness of each proposal at the hands of the publisher, individual contributions are all refereed to the high quality standards of leading journals in the field. Thus, this series provides the research community with well-edited, authoritative reports on developments in the most exciting areas of mathematical and statistical research today.

Sylvia Frühwirth-Schnatter • Angela Bitto
Gregor Kastner • Alexandra Posekany
Editors

Bayesian Statistics from Methods to Models and Applications

Research from BAYSM 2014

Springer

Editors
Sylvia Frühwirth-Schnatter
Institute for Statistics and Mathematics
WU Vienna University of Economics and Business
Vienna, Austria

Angela Bitto
Institute for Statistics and Mathematics
WU Vienna University of Economics and Business
Vienna, Austria

Gregor Kastner
Institute for Statistics and Mathematics
WU Vienna University of Economics and Business
Vienna, Austria

Alexandra Posekany
Institute for Statistics and Mathematics
WU Vienna University of Economics and Business
Vienna, Austria

ISSN 2194-1009 ISSN 2194-1017 (electronic)
Springer Proceedings in Mathematics & Statistics
ISBN 978-3-319-16237-9 ISBN 978-3-319-16238-6 (eBook)
DOI 10.1007/978-3-319-16238-6

Library of Congress Control Number: 2015937960

Springer Cham Heidelberg New York Dordrecht London

Printed on acid-free paper

Springer International Publishing AG Switzerland is part of Springer Science+Business Media (www.springer.com)

The Contribution of Young Researchers to Bayesian Statistics II

Proceedings of BAYSM 2014

Preface

BAYSM 2014—the second Bayesian Young Statisticians Meeting—took place at the WU Vienna University of Economics and Business, Austria, on September 18–19, 2014. The conference was hosted by the Institute for Statistics and Mathematics of the Department of Finance, Accounting and Statistics. It attracted more than 100 participants from 25 different countries spread over five continents.

Following BAYSM 2013, the first meeting of this kind in Milan, Italy, BAYSM 2014 continues to establish a scientific forum for the next generation of researchers in Bayesian statistics. This inspiring scientific meeting provided opportunities for M.S. students, Ph.D. students, postdoctoral scholars, young researchers, and interested parties from the industry to get in touch with the Bayesian community at large, to expand their professional network, to interact with colleagues, and to exchange ideas.

The scientific program reflected the wide variety of fields in which Bayesian methods are currently employed or could be introduced in the future. Three brilliant keynote lectures by Chris Holmes (University of Oxford), Christian Robert (Université Paris-Dauphine), and Mike West (Duke University) were complemented by 24 plenary talks covering the major topics *Dynamic Models*, *Applications*, *Bayesian Nonparametrics*, *Biostatistics*, *Bayesian Methods in Economics*, and *Models and Methods*, as well as a lively poster session with 30 contributions. The presence of numerous "matured" Bayesians, be it keynote speakers, members of the scientific committee, or senior discussants, provided invaluable inspiration for all attendant young researchers. Throughout the whole workshop, participants were able to discuss open questions, received helpful feedback on their current research, and were encouraged to pursue their line of research.

This volume comprises a peer-reviewed selection of young researchers' contributions presented at BAYSM 2014. It is structured in the following way: The first part, entitled *Theory and Methods*, is dedicated to mathematical statistics, model building, and methodological works, demonstrated by examples. The second part, entitled *Applications and Case Studies*, focuses on the applications of complex methods to real-world problems and data. We want to thank all the authors for their excellent contributions to this volume. Thanks are also due to all reviewers for dedicating time and efforts to the improvement of these young researchers' scientific attempts.

We would like to take this opportunity to express our gratitude to all those people who made BAYSM 2014 an outstanding scientific event and an enjoyable experience. We wish to thank our profound keynote speakers, Chris Holmes, Christian Robert, and Mike West for their inspiring talks and their most valued contributions to a lively meeting. Sincere thanks are given to all participants for the high quality of their presentations. Special thanks go to all the senior discussants for their valuable feedback, especially Jesus Crespo Cuaresma, Bettina Grün, Helga Wagner, and the current President of the International Society for Bayesian Analysis, Sonia Petrone. Finally, we are deeply grateful for the outstanding support we received from the organizing committee, chaired by Karin Haupt, Deputy Head of Office FAS D4, the WU Vienna University of Economics and Business, as well as our sponsors, Accenture, Google, ISBA, and UNIQA.

Hosting this meeting was an exciting and most rewarding experience for us, and we are very pleased that BAYSM 2014 could continue the great success of the first meeting. This extraordinary series of scientific meetings for young researchers in Bayesian statistics will be resumed in June 2016 in Florence, Italy, with BAYSM 2016. Further information can be found on the BAYSM websites at baysm2014.wu.ac.at and baysm.org.

Vienna, Austria
December 2014

Sylvia Frühwirth-Schnatter
Angela Bitto
Gregor Kastner
Alexandra Posekany

Organization

Conference Chairs

Sylvia Frühwirth-Schnatter
Angela Bitto
Gregor Kastner
Alexandra Posekany

Scientific Committee

Raffaele Argiento
Angela Bitto
Jesus Crespo Cuaresma
Sylvia Frühwirth-Schnatter
Bettina Grün
Paul Hofmarcher
Francesca Ieva
Gregor Kastner
Ettore Lanzarone
Alexandra Posekany
Laura Vana
Helga Wagner

Contents

Part II Applications and Case Studies

Editors' Biographies

Sylvia Frühwirth-Schnatter is a Professor of Applied Statistics and Econometrics at the Department of Finance, Accounting and Statistics at the WU Vienna University of Economics and Business, Austria. She received her Ph.D. in Mathematics from the Vienna University of Technology in 1988. She has published in many leading journals in applied statistics and econometrics on topics such as Bayesian inference, finite mixture models, Markov switching models, state space models, and their application in economics, finance, and business. In 2014, she became elected member of the Austrian Academy of Science.

Angela Bitto holds a Masters in Mathematics and is currently working on her Ph.D. in Statistics at the Vienna University of Technology. Her research focuses on the Bayesian estimation of sparse time-varying parameter models. Prior to joining the Institute of Statistics and Mathematics at the WU Vienna University of Economics and Business, she worked as a research analyst for the European Central Bank.

Gregor Kastner is an Assistant Professor at the WU Vienna University of Economics and Business and a Lecturer at the University of Applied Sciences in Wiener Neustadt, Austria. He holds Masters in Mathematics, Computer Science, Informatics Management, and Physical Education; in 2014, he received his Ph.D. in Mathematics. Gregor researches the Bayesian modeling of economic time series, in particular the efficient estimation of univariate and high-dimensional stochastic volatility models. His work has been published in leading journals in computational statistics and computer software.

Alexandra Posekany is an Assistant Professor at the Institute of Statistics and Mathematics, WU Vienna University of Economics and Business, Austria. She holds a Ph.D. in Mathematics from the Vienna University of Technology. Her research includes applications of Bayesian analysis in computational biology and econometrics, as well as the development of algorithms and statistical methods in Bayesian computing and big data analysis.

Part I
Theory and Methods

Chapter 1
Bayesian Survival Model Based on Moment Characterization

Julyan Arbel, Antonio Lijoi, and Bernardo Nipoti

Abstract Bayesian nonparametric marginal methods are very popular since they lead to fairly easy implementation due to the formal marginalization of the infinite-dimensional parameter of the model. However, the straightforwardness of these methods also entails some limitations. They typically yield point estimates in the form of posterior expectations, but cannot be used to estimate non-linear functionals of the posterior distribution, such as median, mode or credible intervals. This is particularly relevant in survival analysis where non-linear functionals such as the median survival time play a central role for clinicians and practitioners. The main goal of this paper is to summarize the methodology introduced in (Arbel, Lijoi and Nipoti, Comput. Stat. Data. Anal. 2015) for hazard mixture models in order to draw approximate Bayesian inference on survival functions that is not limited to the posterior mean. In addition, we propose a practical implementation of an R package called **momentify** designed for moment-based density approximation. By means of an extensive simulation study, we thoroughly compare the introduced methodology with standard marginal methods and empirical estimation.

Key words: Bayesian nonparametrics, Completely random measures, Hazard mixture models, Median survival time, Moment-based approximations, Survival analysis

---

J. Arbel (✉)
Collegio Carlo Alberto, Moncalieri, Italy
e-mail: julyan.arbel@carloalberto.org

A. Lijoi
Department of Economics and Management, University of Pavia, Pavia, Italy

Collegio Carlo Alberto, Moncalieri, Italy
e-mail: lijoi@unipv.it

B. Nipoti
Department of Economics and Statistics, University of Torino, Torino, Italy

Collegio Carlo Alberto, Moncalieri, Italy
e-mail: bernardo.nipoti@carloalberto.org

S. Frühwirth-Schnatter et al. (eds.), *Bayesian Statistics from Methods to Models and Applications*, Springer Proceedings in Mathematics & Statistics 126,
DOI 10.1007/978-3-319-16238-6_1

1.1 Introduction

With *marginal methods* in Bayesian nonparametrics we refer to inferential procedures which rely on the integration (or marginalization) of the infinite-dimensional parameter of the model. This marginalization step is typically achieved by means of the so-called Blackwell–MacQueen Pólya urn scheme. We consider the popular example of the Dirichlet process [4] to illustrate the idea. Denote by $\boldsymbol{Y} = (Y_1, \ldots, Y_n)$ an exchangeable sequence of random variables to which we assign as a prior distribution a Dirichlet process with mass parameter M and base measure G_0, that is

$$
\begin{aligned}
Y_i | G &\overset{\text{iid}}{\sim} G, \\
G &\sim DP(M, G_0).
\end{aligned}
$$

The marginal distribution of $\boldsymbol{Y}$, once G has been integrated out, can be derived from the set of predictive distributions for Y_i, given $(Y_1, \ldots, Y_{i-1})$, for each $i = 1, \ldots, n$. In this case, such conditional distributions are linear combinations between the base measure G_0 and the empirical distribution of the conditioning variables and are effectively described through a Pólya urn sampling scheme. Marginal methods have played a major role in the success of Bayesian nonparametrics since the Pólya urn generally leads to ready to use Markov chain Monte Carlo (MCMC) sampling strategies which, furthermore, immediately provide Bayesian point estimators in the form of posterior means. A popular example is offered by mixtures of the Dirichlet process for density estimation; for the implementation, see, e.g., the R package **DPpackage** by Jara et al. [9]. However, the use of marginal methods has important limitations that we wish to address here. Indeed, one easily notes that the posterior estimates provided by marginal methods are not suitably endowed with measures of uncertainty such as posterior credible intervals. Furthermore, using the posterior mean as an estimator is equivalent to choosing a square loss function which does not allow for other types of estimators such as median or mode of the posterior distribution. Finally, marginal methods do not naturally lead to the estimation of non-linear functionals of the distribution of a survival time, such as the median survival time. For a discussion of these limitations, see, e.g., Gelfand and Kottas [5].

The present paper aims at proposing a new procedure that combines closed-form analytical results arising from the application of marginal methods with an approximation of the posterior distribution which makes use of posterior moments. The whole machinery is developed for the estimation of survival functions that are modeled in terms of hazard rate functions. To this end, let F denote the cumulative distribution function (CDF) associated with a probability distribution on $\mathbb{R}^+$. If F is absolutely continuous, then the corresponding survival function and cumulative hazard rate are defined, respectively, by $S(t) = 1 - F(t)$ and $H(t) = -\log(S(t))$, and the hazard rate function is given by $h(t) = -S'(t)/S(t)$. Let us recall that survival analysis has been a very active area of application of Bayesian nonparametric methodology: neutral to the right processes were used by [2] as a prior for the

CDF F, and beta processes by [6] as a prior for the cumulative hazard function H, both benefiting from useful conjugacy properties. Here we specify a prior on the hazard rate h. The most popular example is the gamma process mixture, originally proposed in [3]. More general models have been studied in later work by [10] and [8]. Bayesian inference for these models often relies on a marginal method, see, e.g., [7]. Although quite simple to implement, marginal methods typically yield estimates of the hazard rate, or equivalently of the survival function, only in the form of the posterior mean at a fixed time point. Working along the lines of Arbel et al. [1], we show that a clever use of a moment-based approximation method does provide a relevant upgrade on the type of inference one can draw via marginal sampling schemes. We should stress that the information gathered by marginal methods is not confined to the posterior mean but is actually much richer and, if properly exploited, can lead to a more complete posterior inference.

Let us briefly introduce Bayesian hazard mixture models. Random parameters, such as the hazard rate and survival function, are denoted with a tilde on top, e.g. $\tilde{h}$ and $\tilde{S}$. We endow $\tilde{h}$ with a prior distribution defined by the distribution of the random hazard rate (RHR)

$$\tilde{h}(t) = \int_{\mathbb{Y}} k(t;y)\tilde{\mu}(\mathrm{d}y), \tag{1.1}$$

where $\tilde{\mu}$ is a completely random measure (CRM) on $\mathbb{Y} = \mathbb{R}^+$, and $k(\cdot\,;\cdot)$ denotes a transition kernel on $\mathbb{R}^+ \times \mathbb{Y}$. Under suitable assumption on the CRM $\tilde{\mu}$, we have $\lim_{t\to\infty}\int_0^t \tilde{h}(s)\mathrm{d}s = \infty$ with probability 1. Therefore, we can adopt the following model

$$\begin{aligned} X_i \,|\, \tilde{P} &\overset{\text{iid}}{\sim} \tilde{P} \\ \tilde{P}((t,\infty)) &\overset{\mathrm{d}}{=} \tilde{S}(t) \overset{\mathrm{d}}{=} \exp\left(-\int_0^t \tilde{h}(s)\,\mathrm{d}s\right), \end{aligned} \tag{1.2}$$

for a sequence of (possibly censored) survival data $\boldsymbol{X} = (X_1,\dots,X_n)$. In this setting, [3] characterizes the posterior distribution of the so-called *extended gamma process*: this is obtained when $\tilde{\mu}$ is a gamma CRM and $k(t;y) = \mathbb{1}_{(0,t]}(y)\,\beta(y)$ for some positive right-continuous function $\beta : \mathbb{R}^+ \to \mathbb{R}^+$. The same kind of result is proved in [10] for *weighted gamma processes* corresponding to RHRs obtained when $\tilde{\mu}$ is still a gamma CRM and $k(\cdot\,;\cdot)$ is an arbitrary kernel. Finally, a posterior characterization has been derived by [8] for any CRM $\tilde{\mu}$ and kernel $k(\cdot\,;\cdot)$.

The rest of the paper is organized as follows. In Sect. 1.2, we provide the closed-form expressions for the posterior moments of the survival function. We then show in Sect. 1.3 how to exploit the expression for the moments to approximate the corresponding density function and sample from it. Finally, in Sect. 1.4 we study the performance of our methodology by means of an extensive simulation study with survival data.

1.2 Moments of the Posterior Survival Function

Closed-form expressions for the moments of any order of the posterior survival curve $\tilde{S}(t)$ at any t are provided in Arbel et al. [1]. For a complete account, we recall the result hereafter. We first need to introduce some notation. A useful augmentation suggests introducing latent random variables $\boldsymbol{Y} = (Y_1, \dots, Y_n)$ such that, building upon the posterior characterization derived by [8], we can derive expressions for the posterior moments of the random variable $\tilde{S}(t)$, where t is fixed, conditionally on $\boldsymbol{X}$ and $\boldsymbol{Y}$. To this end, define $K_x(y) = \int_0^x k(s;y)\mathrm{d}s$ and $K_{\boldsymbol{X}}(y) = \sum_{i=1}^n K_{X_i}(y)$. Also, the almost sure discreteness of $\tilde{\mu}$ implies there might be ties among the Y_i's with positive probability. Therefore, we denote the distinct values among $\boldsymbol{Y}$ by $(Y_1^*, \dots, Y_k^*)$, where $k \leq n$, and, for any $j = 1, \dots, k$, we define $C_j = \left\{l : Y_l = Y_j^*\right\}$ and $n_j = \#C_j$. We can now state the following result.

Proposition 1. *Denote by* $\nu(\mathrm{d}s, \mathrm{d}y) = \rho(s)\,\mathrm{d}s\, c\, P_0(\mathrm{d}y)$ *the Lévy intensity of the completely random measure* $\tilde{\mu}$. *Then for every* $t > 0$ *and* $r > 0$,

$$\mathbb{E}[\tilde{S}^r(t)\,|\,\boldsymbol{X}, \boldsymbol{Y}] = \exp\left\{-\int_{\mathbb{R}^+ \times \mathbb{Y}} \left(1 - \mathrm{e}^{-rK_t(y)s}\right) \mathrm{e}^{-K_{\boldsymbol{X}}(y)s} \nu(\mathrm{d}s, \mathrm{d}y)\right\}$$

$$\times \prod_{j=1}^{k} \frac{1}{B_j} \int_{\mathbb{R}^+} \exp\left\{-s\left(rK_t(Y_j^*) + K_{\boldsymbol{X}}(Y_j^*)\right)\right\} s^{n_j} \rho(s)\mathrm{d}s, \qquad (1.3)$$

where $B_j = \int_{\mathbb{R}^+} s^{n_j} \exp\left\{-sK_{\boldsymbol{X}}(Y_j^*)\right\} \rho(s)\mathrm{d}s$, *for* $j = 1, \dots, k$.

For evaluating the posterior moments $\mathbb{E}[\tilde{S}^r(t)\,|\,\boldsymbol{X}]$ by means of Proposition 1, we use a Gibbs sampler which proceeds by alternately sampling, at each iteration $\ell = 1, \dots, L$, from the full conditional distributions of the latent variables $\boldsymbol{Y}$ and the parameters of the model, and evaluating $\mathbb{E}[\tilde{S}^r(t)\,|\,\boldsymbol{X}, \boldsymbol{Y}]^{(\ell)}$ at each step. For an exhaustive description of the posterior sampling and the expression of the full conditional distributions, see Arbel et al. [1]. The remain of the paper is devoted to illustrating how the characterization of the moments provided by Proposition 1 can be used to approximate a density function and, in turn, to carry out Bayesian inference.

1.3 Moment-Based Density Approximation

The aim is to recover the posterior distribution of the random variable $\tilde{S}(t)$ for any fixed t, based on the knowledge of its moments $\mathbb{E}[\tilde{S}^r(t)\,|\,\boldsymbol{X}]$ obtained from Proposition 1. In order to simplify the notation, let us consider a generic continuous random variable S on $[0, 1]$, and denote by f its density, and its raw moments by $\gamma_r = \mathbb{E}[S^r]$, with $r \in \mathbb{N}$. Recovering f from the explicit knowledge of its moments γ_r is a classical problem in probability and statistics that has received great attention in the literature, see, e.g., [11] and the references and motivating applications therein. A very general approach relies on the basis of Jacobi polynomials $(G_i(s) = \sum_{r=0}^{i} G_{i,r}s^r)_{i\geq 1}$. They constitute a broad class which includes, among others, Legendre and Chebyshev polynomials, and which is well suited for the expansion of densities with compact support [see 11]. Any univariate density f supported on $[0,1]$ can be uniquely decomposed on such a basis and therefore there is a unique sequence of real numbers $(\lambda_i)_{i\geq 0}$ such that $f(s) = w_{a,b}(s)\sum_{i=0}^{\infty}\lambda_i G_i(s)$ where $w_{a,b}(s) = s^{a-1}(1-s)^{b-1}$ is named the *weight function* of the basis and is proportional to a beta density in the case of Jacobi polynomials. From the evaluation of $\int_0^1 f(s)\,G_i(s)\,\mathrm{d}s$ it follows that each λ_i coincides with a linear combination of the first i moments of S, specifically $\lambda_i = \sum_{r=0}^{i} G_{i,r}\gamma_r$. Then, the polynomial approximation method consists in truncating the representation of f in the Jacobi basis at a given level $i = N$. This procedure leads to a methodology that makes use only of the first N moments and provides the approximation

$$f_N(s) = w_{a,b}(s)\sum_{i=0}^{N}\left(\sum_{r=0}^{i} G_{i,r}\mu_r\right) G_i(s). \tag{1.4}$$

It is important to stress that the polynomial approximation (1.4) is not necessarily a density as it might fail to be positive or to integrate to 1. In order to overcome this problem, we consider the density π proportional to the positive part of (1.4) defined by $\pi(s) \propto \max(f_N(s), 0)$. We resort to the *rejection sampler* for sampling from π. This is a method for drawing independently from a distribution proportional to a given non-negative function, that exempts us from computing the normalizing constant corresponding to π. More precisely, the method requires to pick a proposal distribution p for which there exists a positive constant M such that $\pi \leq Mp$. A natural choice for p is the beta distribution proportional to the weight function $w_{a,b}$. Approximation (1.4) and the rejection sampler were implemented in R. For the purpose of the paper, we have wrapped up the corresponding code in an R package called **momentify**.[1] In Sects 1.3.1 and 1.3.2 we briefly describe the package implementation and give a simple working example.

1.3.1 Package Implementation

The major function in package **momentify** is called `momentify` and allows for (i) approximating a density based on its moments, and (ii) sampling from

[1] The **momentify** package can be downloaded from the first author's webpage http://www.crest.fr/pagesperso.php?user=3130.

this approximate distribution by using the rejection sampler. The synopsis of the function, together with default values, is given by

```
momentify(moments, N_moments = length(moments),
 N_sim = 1000, xgrid = seq(0, 1, length = 200))
```

The only required argument is `moments`, a d-dimensional vector, with $d \geq 2$, composed by the values of the first d consecutive raw moments. The remaining arguments are optional: `N_moments` corresponds to N, the number of moments to be used (where $N \leq d$), `N_sim` is the size of the sample obtained by the rejection sampler, and `xgrid` denotes the grid on which the density is to be approximated.

The function returns a list, say `res`, with the following components: `xgrid`, defined in argument, `approx_density`, the approximated density evaluated on `xgrid`, and `psample`, the sample obtained from `approx_density` by the reject algorithm. The class of the output list `res` is called `momentify`. For visualizing the output `res`, two method functions can be readily applied to this class, namely `plot(res, ...)` and `hist(res, ...)`.

1.3.2 Simulated Example

We assess now the quality of this approximation procedure on a particular example by means of a practical implementation of the **momentify** package. We specify the distribution of the random variable S by a mixture, with weights of 1/2, of beta distributions of parameters $(a,b) = (3,5)$ and $(c,d) = (10,3)$. The raw moments of any order of S can be explicitly evaluated by

$$\gamma_r = \mathbb{E}[S^r] = \frac{a_{(r)}}{(a+b)_{(r)}} + \frac{c_{(r)}}{(c+d)_{(r)}},$$

where $x_{(r)} = \Gamma(x+r)/\Gamma(x)$. As described above, given a vector of N moments $(\gamma_1, \dots, \gamma_N)$, the introduced package allows us to approximately evaluate the density (1.4) and, in turn, to compare it with the true density. The corresponding code for $N = 2, \dots, 10$ is the following:

```
rfun=function(n){bin=rbinom(n,1,.5)
     bin*rbeta(n,3,5)+(1-bin)*rbeta(n,10,3)}
true_density=function(n){.5*dbeta(n,3,5)+
                         .5*dbeta(n,10,3)}
sim_data = rfun(10^5)
moments = mean(sim_data)
for (i in 2:10){
  moments = c(moments,mean(sim_data^i))
  res = momentify(moments = moments)
  plot(res, main = paste("N =",i))
  curve(true_density(x),add=TRUE, col = "red")
}
```

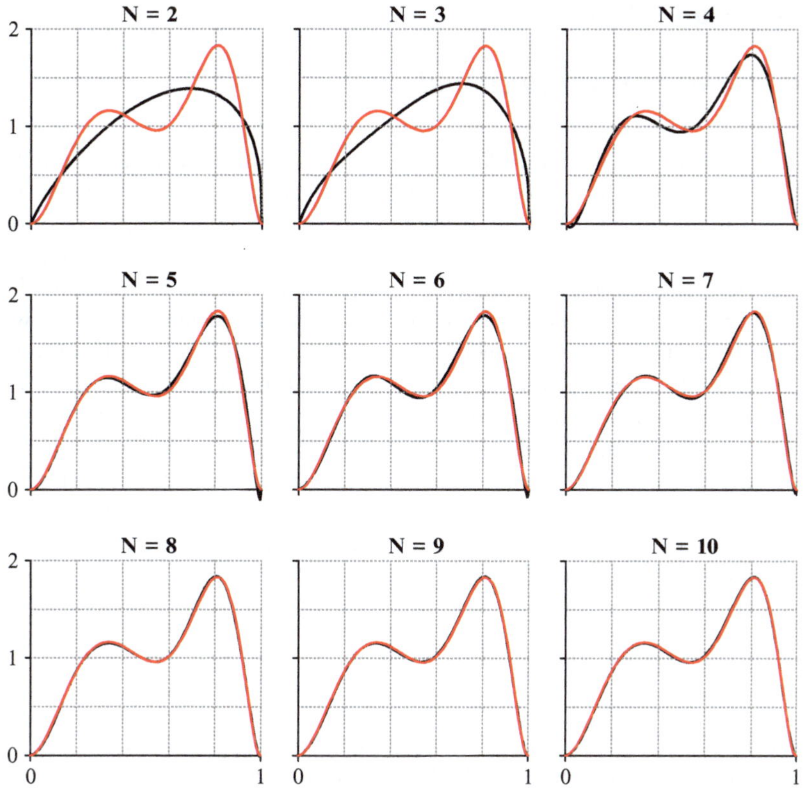

Fig. 1.1 Output of **momentify** R package. True density f of S (in *red*) and approximated density f_N (in *black*) involving an increasing number of moments, from $N = 2$ (*top left*) to $N = 10$ (*bottom right*)

The graphical output is given in Fig. 1.1. We can see that four moments are needed in order to capture the two modes of the distribution, although coarsely. From seven moments onward, the fit is very good since the two curves are hardly distinguishable. Following this example as well as other investigations not reported here, we choose, as a rule of thumb, to work with $N = 10$ moments. A more elaborated numerical study is presented in Arbel et al. [1] in the context of survival analysis.

1.4 Bayesian Inference

1.4.1 Estimation of Functionals of $\tilde{S}$

Given a sample of survival times $\boldsymbol{X} = \{X_1, \dots, X_n\}$, we estimate the first N moments of the posterior distribution of $\tilde{S}(t)$, for t on a grid of q equally spaced points $\{t_1, \dots, t_q\}$ in an interval $[0, M]$ by using the Gibbs sampler succinctly described in Sect. 1.2. We then exploit the estimated moments to sample from an approximation of the posterior distribution of $\tilde{S}(t_i)$ for $i = 1, \dots, q$ according to the methodology set forth in Sect. 1.3. This allows us to carry out Bayesian inference, with a focus on the estimation of the median survival time and, for any given t in the grid, of credible intervals for $\tilde{S}(t)$. The same approach can be easily used to estimate the posterior median and mode of $\tilde{S}(t)$ at any given t, and, in line of principle, any functional of interest. Let us first consider the median survival time that we denote by m. The identity for the cumulative distribution function of m, $\mathbb{P}(m \le t | \boldsymbol{X}) = \mathbb{P}\big(\tilde{S}(t) \le 1/2 | \boldsymbol{X}\big)$, allows us to evaluate the CDF of m at each time point t_i as $c_i = \mathbb{P}\big(\tilde{S}(t_i) \le 1/2 | \boldsymbol{X}\big)$. Then, we can estimate the median survival time m by means of the following approximation:

$$\hat{m} = \mathbb{E}_{\boldsymbol{X}}[m] = \int_0^\infty \mathbb{P}[m > t | \boldsymbol{X}] \, \mathrm{d}t \approx \frac{M}{q-1} \sum_{i=1}^{q} (1 - c_i), \tag{1.5}$$

where the subscript $\boldsymbol{X}$ in $\mathbb{E}_{\boldsymbol{X}}[m]$ indicates that the integral is with respect to the distribution of $\tilde{S}(\cdot)$ conditional to $\boldsymbol{X}$. Moreover, the sequence $(c_i)_{i=1}^q$ can be used to devise credible intervals for the median survival time. Similarly, the posterior samples generated by the rejection sampler can be easily used to devise, t-by-t, credible intervals for $\tilde{S}(t)$ or to estimate other functionals that convey meaningful information such as the posterior mode and median. In Sect. 1.4.2, we apply this methodology in a study involving simulated survival data where we compare the performance of the moment-based methodology with standard marginal methods.

1.4.2 Applications

For the purpose of illustration, we complete the model specification by assuming a Dykstra and Laud type of kernel $k(t; y) = \mathbb{1}_{(0,t]}(y)\beta$, for some constant $\beta > 0$, a gamma CRM $\tilde{\mu}$ and an exponential base measure P_0 with rate parameter 3. Moreover, for the hyperparameters c and β we choose independent gamma prior distributions with shape parameter 1 and rate parameter $1/3$. Then, we consider three samples $\boldsymbol{X} = (X_1, \dots, X_n)$ of size $n = 20, 100, 500$ from a Weibull distribution of parameters $(2, 2)$ whose survival function is $S_0(t) = \exp(-t^2/4)$. We set $M = 6$ (the largest observation in the samples is 5.20) and $q = 50$ for the analysis of each sample. We approximately evaluate, t-by-t, the posterior distribution of $\tilde{S}(t)$ together

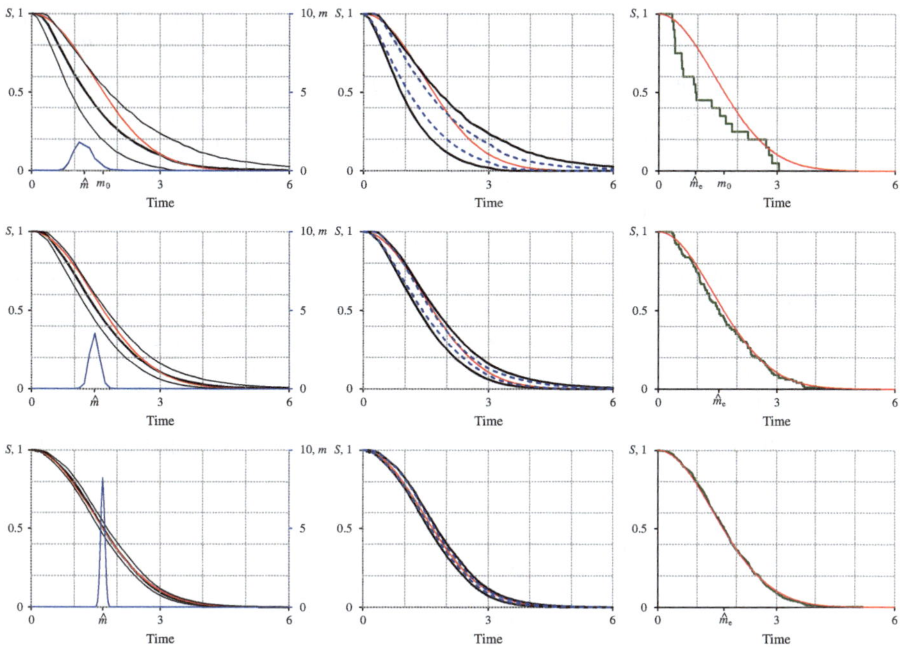

Fig. 1.2 The true survival function $S_0(t)$ is the *red line* in all plots. *Bottom row*: estimated posterior mean (*black solid line*) with 95 % credible intervals for $\tilde{S}(t)$ (*black thin lines*); in *blue* the posterior distribution of the median survival time *m*. *Middle row*: comparison of the 95 % credible interval (*black line*) with the marginal interval (*dashed blue line*). *Top row*: Kaplan–Meier estimate (*green line*). Sample size n=20 (*top row*), n=100 (*middle row*), n=500 (*bottom row*)

with the posterior distribution of the median survival time m. By inspecting the bottom row of Fig. 1.2, we can appreciate that the estimated credible intervals for $\tilde{S}(t)$ contain the true survival function. Moreover, the posterior distribution of the median survival time m (blue curve) is nicely concentrated around the true value m_0. When relying on marginal methods, the most natural choice for quantifying the uncertainty of posterior estimates consists of considering the quantile intervals corresponding to the output of the Gibbs sampler, that we refer to as *marginal intervals*. This leads to considering, for any fixed t, the interval whose lower and upper extremes are the quantiles of order, e.g. 0.025 and 0.975, respectively, of the sample of posterior means $\big(\mathbb{E}[\tilde{S}(t)\,|\,\boldsymbol{X},\boldsymbol{Y}]^{(\ell)}\big)_{\ell=1,\dots,L}$ obtained, conditional on $\boldsymbol{Y}$, by the Gibbs sampler described in Sect. 1.2. In the middle row of Fig. 1.2 we have compared the estimated 95 % credible intervals for $\tilde{S}(t)$ (black) and the marginal intervals corresponding to the output of the Gibbs sampler (dashed blue). In this example, the credible intervals in general contain the true survival function $S_0(t)$, while this does not hold for the marginal intervals. This fact suggests that the marginal method tends to underestimate the uncertainty associated with the posterior estimates, and can be explained by observing that, since the underlying CRM is marginalized out, the intervals arising from the Gibbs sampler output capture only the variability of the posterior mean that can be traced back to the marginalization with respect to latent variables $\boldsymbol{Y}$ and the parameters (c,β). As a result, especially for a small sample size, the uncertainty detected by the marginal method leads to marginal intervals that can be significantly narrower than the actual posterior credible intervals that we approximate through the moment-based approach. The Kaplan–Meier estimates of $\tilde{S}(t)$ are plotted on the top row of Fig. 1.2.

On the one hand, as described in Sect. 1.4.1, the moment-based approach enables us to approximate the posterior distribution of the median survival time m (in blue in the bottom row of Fig. 1.2). This, in turn, can be used to derive sensible credible intervals for m. On the other hand, when relying on marginal methods, the posterior of the median survival time is not available *per se*. However, in the same way as we defined *marginal intervals* in place of credible intervals for the survival function $\tilde{S}(t)$, for every t_i the Gibbs sample $\big(\mathbb{E}[\tilde{S}(t_i)\,|\,\boldsymbol{X},\boldsymbol{Y}]^{(\ell)}\big)_{\ell=1,\dots,L}$ can be used as a proxy of a posterior sample for $\tilde{S}(t_i)$ in order to provide the following approximation of the CDF of m:

$$\mathbb{P}\left(m \le t|\boldsymbol{X}\right) \approx 1/2|\boldsymbol{X}) = \frac{1}{L}\#\{\ell : \mathbb{E}[\tilde{S}(t)\,|\,\boldsymbol{X},\boldsymbol{Y}]^{(\ell)} \le 1/2\}. \tag{1.6}$$

As in (1.5), an estimator for the median survival time can be obtained as the mean of the distribution whose CDF is given in (1.6). We call such an estimator $\hat{m}_m$ to denote the fact that it is obtained by means of a marginal method. Similarly, from (1.6), marginal intervals for $\hat{m}_m$ can be derived as described in Sect. 1.4.1. Finally, we denote by $\hat{m}_e$ the empirical estimator of m and by $m_0 = 2\sqrt{\log 2} \approx 1.665$ the true median survival time. We summarize the estimates we obtained for the median survival time m in Fig. 1.3 and in Table 1.1. For all the sample sizes considered, the credible intervals for $\hat{m}$ contain the true value. Moreover, as expected, when n grows they shrink toward m_0: for instance, the length of the interval reduces from 0.92 to 0.20, when the sample size n increases from 20 to 500. As observed for

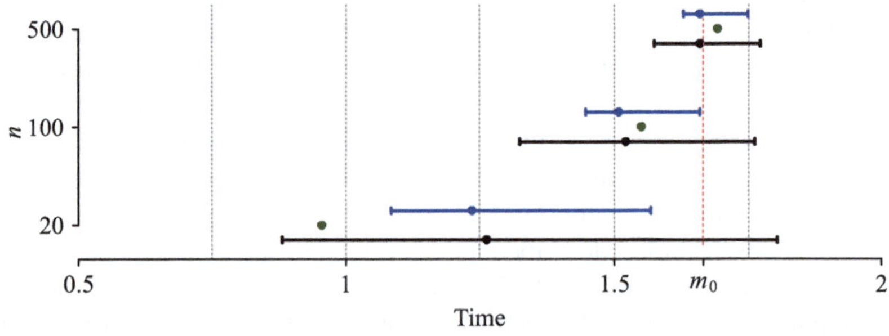

Fig. 1.3 Comparison of credible intervals for the median survival time m obtained with the moment-based approach (*black line*, below for each n) and marginal intervals (*blue line*, above for each n), for varying sample size n. The *dots* indicate the estimators ($\hat{m}$ in *black*, $\hat{m}_m$ in *blue* and $\hat{m}_e$ in *green*). The true median $m_0 = 2\sqrt{\log 2} \approx 1.665$ is indicated by the *vertical red dashed line*

Table 1.1 Comparison of the median survival time estimated by means of the moment-based method, $\hat{m}$, by means of the marginal method, $\hat{m}_m$, and the empirical median survival time $\hat{m}_e$, for different sample sizes n

	Moment-based method			Marginal			Empirical	
n	$\hat{m}$	$\lvert\hat{m}-m_0\rvert$	CI	$\hat{m}_m$	$\lvert\hat{m}_m-m_0\rvert$	CI_m	$\hat{m}_e$	$\lvert\hat{m}_e-m_0\rvert$
20	1.26	0.40	0.88–1.80	1.24	0.43	1.09–1.57	0.95	0.71
100	1.52	0.14	1.33–1.76	1.51	0.16	1.45–1.66	1.55	0.12
500	1.66	0.01	1.57–1.77	1.66	0.01	1.63–1.75	1.69	0.03

For the moment-based estimation we show $\hat{m}$, the absolute error $|\hat{m}-m_0|$ and 95 % credible interval (CI); for the marginal method, we show $\hat{m}_m$, the absolute error $|\hat{m}_m-m_0|$ and the 95 % marginal interval (CI_m); the last two columns show the empirical estimate $\hat{m}_e$ and the corresponding absolute error $|\hat{m}_e-m_0|$. The true median survival time is $m_0 = 2\sqrt{\log 2} \approx 1.665$

the marginal intervals $\tilde{S}(t)$ at a given t, the marginal intervals for $\hat{m}_m$ obtained with the marginal method and described in Equation (1.6) are in general narrower than the credible intervals obtained by the moment-based approach. Moreover, in this example, they contain the true m_0 only for $n = 500$. This observation suggests that the use of intervals produced by marginal methods as proxies for posterior credible intervals should be avoided, especially for small sample sizes.

Acknowledgements J. Arbel and A. Lijoi are supported by the European Research Council (ERC) through StG "N-BNP" 306406.

References

[1] Arbel, J., Lijoi, A., Nipoti, B.: Full Bayesian inference with hazard mixture models. To appear in Comput. Stat. Data. Anal. (2015) http://dx.doi.org/10.1016/j.csda.2014.12.003

[2] Doksum, K.: Tailfree and neutral random probabilities and their posterior distributions. Ann. Probab. **2**(2), 183–201 (1974)

[3] Dykstra, R., Laud, P.: A Bayesian nonparametric approach to reliability. Ann. Stat. **9**(2), 356–367 (1981)

[4] Ferguson, T.: A Bayesian analysis of some nonparametric problems. Ann. Stat. **1**(2), 209–230 (1973)

[5] Gelfand, A.E., Kottas, A.: A computational approach for full nonparametric Bayesian inference under Dirichlet process mixture models. J. Comput. Graph. Stat. **11**(2), 289–305 (2002)

[6] Hjort, N.: Nonparametric Bayes estimators based on beta processes in models for life history data. Ann. Stat. **18**(3), 1259–1294 (1990)

[7] Ishwaran, H., James, L.: Computational methods for multiplicative intensity models using weighted gamma processes: proportional hazards, marked point processes, and panel count data. J. Am. Stat. Assoc. **99**(465), 175–190 (2004)

[8] James, L.: Bayesian Poisson process partition calculus with an application to Bayesian Lévy moving averages. Ann. Stat. **33**(4), 1771–1799 (2005)
[9] Jara, A., Hanson, T., Quintana, F., Müller, P., Rosner, G.: DPpackage: Bayesian non-and semi-parametric modelling in R. J. Stat. Softw. **40**(5), 1–30 (2011)
[10] Lo, A., Weng, C.: On a class of Bayesian nonparametric estimates. II. Hazard rate estimates. Ann. I Stat. Math. **41**(2), 227–245 (1989)
[11] Provost, S.B.: Moment-based density approximants. Math. J. **9**(4), 727–756 (2005)

Chapter 2
A New Finite Approximation for the NGG Mixture Model: An Application to Density Estimation

Ilaria Bianchini

Abstract A new class of random probability measures, approximating the well-known normalized generalized gamma (NGG) process, is defined. The new process is built from the representation of the NGG process as a discrete measure, where the weights are obtained by normalization of points of a Poisson process larger than a threshold ε. Consequently, the new process has an as surely finite number of location points. This process is then considered as the mixing measure in a mixture model for density estimation; we apply it to the popular Galaxy dataset. Moreover, we perform some robustness analysis to investigate the effect of the choice of the hyperparameters.

Key words: Bayesian nonparametric mixture models, A-priori truncation method, Normalized generalized gamma process

2.1 Introduction to Bayesian Nonparametric Mixture Models

In this first section we deal with the problem of density estimation from a Bayesian nonparametric point of view. The nonparametric approach is very useful because it allows a rich class of models for the data, considering infinite dimensional families of probability models. Priors on such families are known as nonparametric Bayesian priors and prevent misleading decisions and inference that may result for a parametric approach, which requires a strong assumption about the investigated

---

I. Bianchini (✉)
Department of Mathematics, Politecnico di Milano, Milan, Italy

Istituto di Matematica Applicata e Tecnologie Informatiche (IMATI), National Research Council, Milan, Italy
e-mail: ilaria.bianchini@polimi.it

S. Frühwirth-Schnatter et al. (eds.), *Bayesian Statistics from Methods to Models and Applications*, Springer Proceedings in Mathematics & Statistics 126,
DOI 10.1007/978-3-319-16238-6_2

phenomenon, cf. [11]. We will see how a particularly flexible class of nonparametric priors within the family of normalized random measures with independent increments (NRMI) can be applied for density estimation problems.

Mixture models provide a statistical framework for modeling a collection of continuous observations $(X_1,\dots,X_n)$ where each measurement is supposed to arise from one of k possible unknown groups and each group is modeled by a density from a suitable parametric family.

This model is usually represented hierarchically in terms of a collection of independent and identically distributed latent random variables $(\theta_1,\dots,\theta_n)$ as follows:

$$\begin{cases} X_i|\theta_i \stackrel{ind}{\sim} K(\cdot|\theta_i), & i=1,\dots,n, \\ \theta_i|P \stackrel{iid}{\sim} P, & i=1,\dots,n, \\ P \sim Q, & \end{cases} \tag{2.1}$$

where Q denotes the nonparametric prior distribution and $K(\cdot|\theta)$ is a probability density function parameterized by the latent random variable θ.

Model (2.1) is equivalent to assume $X_1,\dots,X_n$ i.i.d. according to a probability density that is a mixture of kernel functions:

$$X_1,\dots,X_n \stackrel{iid}{\sim} f(x) = \int_\Theta K(x|\theta)P(d\theta), \tag{2.2}$$

where P is called mixing measure. Note that if Q selects discrete probability measures, P is almost surely (a.s.) discrete and the mixture model can be written as a sum with a countably infinite number of components:

$$f(x) = \sum_{j=1}^{\infty} p_j K(x|\theta_j),$$

where the weights $(p_j)_{j\geqslant 1}$ represent the relative frequencies of the groups in the population indexed by θ_j. This approach provides a flexible model for clustering items in a hierarchical setting without the necessity to specify in advance the exact number of clusters; therefore, it can also be adopted in cluster analysis. In the next section, the normalized generalized gamma (NGG) prior is introduced, starting from its construction via normalization of a discrete random measure. As we will see, it is very flexible and still mathematically tractable at the same time, making it a suitable choice for Q in the mixture model.

2.2 The NGG Process

Here, we briefly recall how to build a normalized random measure with independent increments (for an in-depth study, see Chapter 8 of [8]). Consider a (a.s.) discrete random measure $\mu(\cdot)$: it can be expressed as an infinite weighted sum of degenerate measures:

$$\mu(\cdot) = \sum_{i \geq 1} J_i \delta_{\tau_i}(\cdot). \tag{2.3}$$

The random elements $(J_i, \tau_i)_{i \geqslant 1}$ are the points of a Poisson process on $(\mathbb{R}^+, \mathbb{X})$ with mean measure ν that satisfies the following conditions:

$$\int_{(0,1)} s\nu(ds, \mathbb{X}) < \infty, \qquad \nu([1, \infty) \times \mathbb{X}) < \infty. \tag{2.4}$$

This construction produces the most general completely random measure (CRM) without fixed atoms and non-random measure parts: it selects discrete measures almost surely.

An important property which one could impose on a CRM is homogeneity, i.e. the underlying mean measure factorizes. Let P_0 be a non-atomic and σ-finite probability measure on $\mathbb{X}$: if $\nu(ds, dx) = \rho(ds)P_0(dx)$, for some measure ρ on $\mathbb{R}^+$, we call μ *homogeneous*: in this case, the jumps in the representation (2.3) are independent from the locations.

The sequence $(J_i)_{i \geq 1}$ represents the jumps controlled by the kernel ρ and $(\tau_i)_{i \geq 1}$ are the locations of the jumps determined by the measure P_0 on $\mathbb{X}$. Since μ is a discrete random measure almost surely, it is straightforward to build a discrete random probability measure by the normalization procedure, which yields NRMIs, first introduced by [14].

Obviously the procedure is well defined only if the total mass of the measure $T := \mu(\mathbb{X})$ is positive and finite almost surely:

$$\mathbb{P}(0 < T < \infty) = 1.$$

This requirement is satisfied if the measure ρ (in the homogeneous case) is such that

$$\int_{\mathbb{R}^+} \rho(ds) = \infty \quad \forall x \in \mathbb{X}. \tag{2.5}$$

This means that the jumps of the process form a dense set in $(0, \infty)$. However, since the second condition in (2.4) must hold, it turns out that infinite points of the Poisson process are very small. In fact, we find that the integral of intensity ρ over $\mathbb{R}^+$ is infinite while the subinterval over $[0, \infty)$ is finite. Now, we define an NRMI $P(\cdot)$ as $\mu(\cdot)/T$. It is important to highlight that NRMIs select, almost surely, discrete distributions, such that P admits a series representation as

$$P = \sum_{j \geq 1} p_j \delta_{\tau_j}, \tag{2.6}$$

where $p_j = J_j/T \; \forall j \geqslant 1$, where the weights J_j are those in (2.3).

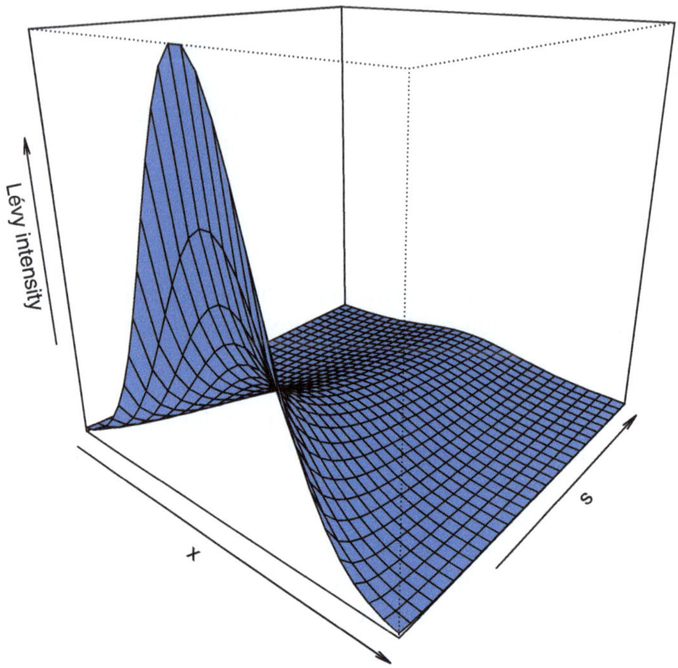

Fig. 2.1 Example of intensity measure $\nu(ds,dx) = 1/\Gamma(1-\sigma)e^{-s}s^{-1-\sigma}dsP_0(dx)$ where $\sigma = 0.1$ and P_0 is Gaussian with mean 0 and variance 1

The NRMI addressed here is the NGG process. As stated in [1], a generalized gamma measure is an NRMI μ with intensity measure equal to

$$\nu(A \times B) = P_0(B) \int_A \rho(ds), \qquad A \in \mathscr{B}(\mathbb{R}^+), B \in \mathscr{B}(\mathbb{X})$$

where

$$\rho(ds) = \frac{\kappa}{\Gamma(1-\sigma)} s^{-1-\sigma} e^{-s\omega} ds, \qquad s > 0. \tag{2.7}$$

Figure 2.1 displays $\nu(s,x)$, where $\omega = \kappa = 1$, $\sigma = 0.1$ and P_0 is Gaussian with mean 0 and variance 1. It is straightforward to define the homogeneous random probability measure $P(\cdot) = \mu(\cdot)/T$ as in (2.6), by the name of NGG process

$$P \sim NGG(\sigma, \kappa, \omega, P_0),$$

with parameters $(\sigma, \kappa, \omega, P_0)$, where $0 \leqslant \sigma \leqslant 1$, $\omega \geq 0$, $\kappa \geq 0$. Within this wide class of priors one finds the following special cases:

1. The Dirichlet process $DP(\kappa, P_0)$ which is an $NGG(0, \kappa, P_0)$ process;
2. The normalized inverse Gaussian process that corresponds to a $NGG(1/2, \kappa, P_0)$.

One could wonder why to choose this process instead of using directly the popular Dirichlet process. The main reason lies in the greater flexibility of the clustering behavior, achieved by the additional parameter, σ, which tunes the variance of the number of distinct observations in a sample from P (if σ increases, the variance increases too; see, for instance, [9]).

2.3 The ε-NGG Approximation

The model we are going to approximate in this section is the so-called NGG mixture model,

$$\begin{cases} X_i|\theta_i \overset{ind}{\sim} K(\cdot|\theta_i), & i=1,\dots,n, \\ \theta_i|P \overset{iid}{\sim} P, & i=1,\dots,n, \\ P \sim NGG(\sigma,\kappa,\omega,P_0). \end{cases} \tag{2.8}$$

From now on, we will consider kernels $K(\cdot|\theta)$ defined on $\mathbb{X} \subseteq \mathbb{R}^p$, where p represents the dimension of the data, and the prior NGG is defined on $\Theta \subseteq \mathbb{R}^m$, the space of the parameters of the kernel. For instance, if K is the univariate Gaussian distribution, $N(\mu, \sigma^2)$, the latent variable θ could be the couple (μ, σ^2), hence $\Theta = (\mathbb{R} \times \mathbb{R}^+)$. The main problem when dealing with nonparametric mixture models is the presence of an infinite dimensional parameter P, which makes these models computationally difficult to handle.

In the literature, one can find two ways to tackle this problem, namely marginal and conditional methods: on the one hand, the first ones integrate out the infinite dimensional parameter, leading to generalized Polya urn schemes (see, for instance, [10] and [12]). This approach has one main limitation: We cannot obtain information about the latent variables, since the posterior inference involves only the predictive distribution $f(X_{n+1}|X_1, X_2, \dots, X_n)$. On the other hand, conditional methods build a Gibbs sampler which does not integrate out the nonparametric mixing measure but update it as a part of the algorithm itself. The reference papers on conditional algorithms for Dirichlet process mixtures are the retrospective sampler of [13] and the slice sampler of [15] (extended in the more general NRMI case in [5]). Conditional methods can also be based on truncation of the sum defining the mixing measure P in (2.6): it can be performed both a-posteriori, as in [6] and [1], or a-priori, as in [7] and [4]. The driving motivation for using conditional methods is that they provide a "full Bayesian analysis," i.e. it is possible to estimate either posterior mean functional or linear and nonlinear functionals, such as quantiles.

The proposed method is based on an *a-priori truncation* of P: in particular, we consider only jumps greater than a threshold $\varepsilon > 0$, which turns out to control the approximation to the infinite dimensional prior: conditionally on ε, only a finite number of jumps has to be considered, hence we resorted to a finite dimensional problem. In particular, the number of jumps J_j greater than a threshold value ε is $N_\varepsilon + 1$, where N_ε is a random variable distributed as

$$N_\varepsilon \sim Poisson(\Lambda_\varepsilon), \qquad \Lambda_\varepsilon = \int_\varepsilon^\infty \rho(ds) = \frac{\kappa\omega^\sigma}{\Gamma(1-\sigma)}\Gamma(-\sigma,\omega\varepsilon),$$

so that its expectation increases as ε decreases. Furthermore, the jumps $(J_0, J_1, \ldots, J_{N_\varepsilon})$ turn out to be i.i.d. from

$$\rho_\varepsilon(s) = \frac{\rho(s)}{\Lambda_\varepsilon}\mathbb{1}_{(\varepsilon,\infty)}(s) = \frac{1}{\omega^\sigma\Gamma(-\sigma,\omega\varepsilon)}s^{-\sigma-1}e^{-\omega s}\mathbb{1}_{(\varepsilon,\infty)}(s).$$

We consider location points $(\tau_0, \tau_1, \ldots, \tau_{N_\varepsilon})$ i.i.d. from the base measure P_0 and define the following discrete (a.s.) random probability measure on Θ:

$$P_\varepsilon(\cdot) = \sum_{j=0}^{N_\varepsilon} \frac{J_j}{T_\varepsilon}\delta_{\tau_j}(\cdot) \tag{2.9}$$

where $T_\varepsilon = \sum_{j=0}^{N_\varepsilon} J_j$. P_ε in (2.9) is denoted as ε-$NGG(\sigma,\kappa,\ \omega,\ P_0)$ process. This process can be seen as an approximated version of the NGG process of Sect. 2.2, provided that ε is small, since the convergence to the NGG process holds true provided that ε tends to 0. The main advantage compared to the corresponding NGG is that in this case the sum defining P_ε is finite: We moved from an infinite dimensional process to a finite dimensional one, which eventually (when ε assumes a very small value) approximates the NGG.

The mixture model we are going to consider can be expressed as follows:

$$\begin{cases} X_1, \ldots, X_n | \theta_1, \ldots, \theta_n \sim \prod_{i=1}^n K(X_i|\theta_i), \\ \theta_1, \ldots, \theta_n | P_\varepsilon \sim P_\varepsilon \text{ i.i.d.}, \\ P_\varepsilon \sim \varepsilon\text{-}NGG(\sigma,\kappa,\omega,P_0), \\ \varepsilon, \sigma, \kappa \sim \pi(\varepsilon,\sigma,\kappa). \end{cases}$$

It can be either considered as an approximation of the NGG mixture model (2.8) or as a separate model when ε is random. In the latter case, we let data "drive" the degree of approximation and the model can be significantly different with respect to its nonparametric counterpart, because ε may assume relatively large values.

Before proceeding to the application of Sect. 2.4, it is useful to remember that the Bayesian estimate of the true density is

$$f_{X_{n+1}}(x|X_1,\ldots,X_n) = \int \sum_{j=0}^{N_\varepsilon} \frac{J_j}{T_\varepsilon} K(x|\tau_j)\mathscr{L}(d\varepsilon, d\sigma, d\kappa, dP|X_1,\ldots,X_n)$$

which will be estimated through Monte Carlo methods.

A more detailed description of the ε-NGG mixture model, providing also a proof of convergence and an MCMC algorithm to sample from the posterior distribution of the model, can be found in [2].

2.4 An Application to Density Estimation for the Galaxy Data

In this section, we apply the model proposed in Sect. 2.3 to a very popular dataset in the literature, the Galaxy dataset, exploiting the Gibbs sampler scheme of [2]. These data are observed velocities of $n = 82$ different galaxies, belonging to six well-separated conic sections of space. Specifically, we use Gaussian kernel densities $K(x|\theta) = N(x|\mu, \sigma^2)$. Hence, P_0, the parameter of the nonparametric prior, is a normal inverse-gamma distribution,

$$N\left(\mu|\bar{X}, \frac{\sigma^2}{0.01}\right) IG\left(\sigma^2|2,1\right),$$

where $\bar{X}$ stands for the sample mean, 20.83. This set of hyperparameters, first proposed by [3], is standard in the literature.

We perform a robustness analysis through a lot of experiments which highlight the relationship between the posterior estimates and the prior choice of the parameters. In fact, the choice of a value (or a prior in the random case) for these parameters is the most complicated part of the model, since it strongly influences the posterior inference.

Here, we present some results corresponding to different sets of hyperparameters: we report in Table 2.1 nine combinations of (σ, κ) together with three values for the a-priori expected values for the number of groups K_n, namely $\{3,5,20\}$, that we used for our experiments.

Obviously, as mentioned in Sect. 2.2, as σ increases, the variance of K_n increases. In addition, we consider three different priors for ε, in order to study their influence on posterior inference. In what follows, we call (A) the case where the prior is degenerate on a value, i.e. $\varepsilon = 10^{-6}$, (B) where $\varepsilon \sim Unif(0,0.1)$ and (C)

Table 2.1 Combinations of parameters (σ, κ) chosen for the numerical examples: we selected three different couples for each prior mean number of groups in the data

Index	$\mathbb{E}(K_n)$	σ	κ
1	3	0.001	0.45
2	3	0.1	0.25
3	3	0.2	0.05
4	5	0.001	1.0
5	5	0.2	0.35
6	5	0.3	0.09
7	20	0.2	5.0
8	20	0.4	2.2
9	20	0.6	0.3

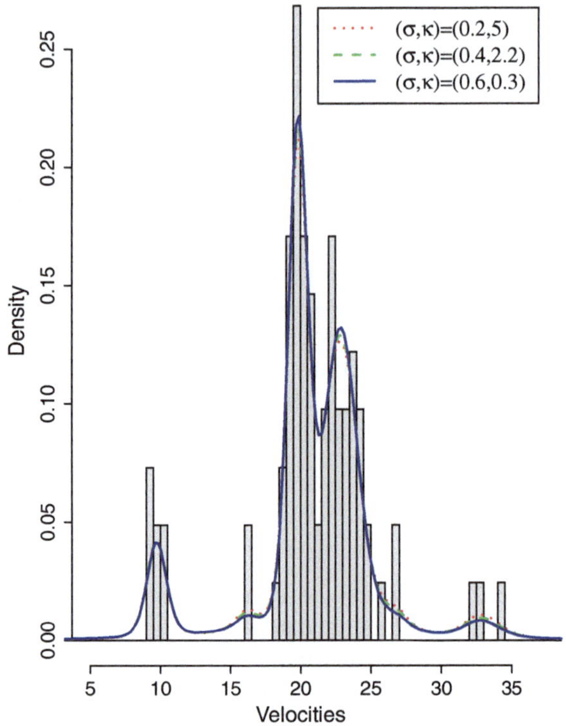

Fig. 2.2 Density estimates for test cases $A7$, $A8$, and $A9$

where ε is a $Beta(0.69, 2.06)$ scaled to the interval $(0, \delta = 0.1)$. In case (C), we chose an informative prior for ε (with mean 0.25δ and variance $0.05\delta^2$) which is concentrated over very small values, since our goal is to approximate the NGG mixture model. Overall, we will have 27 test cases named $A1, \ldots, A9$, $B1, \ldots, B9$, $C1, \ldots, C9$.

Figure 2.2 shows the posterior estimates in test cases $A7$, $A8$, and $A9$, proving reasonable density estimates. We notice that there are only slight differences between the various density estimates, indicating robustness of the model. Figure 2.3 demonstrates that, when σ assumes larger values, the posterior distributions of K_n spread to a larger range of possible values. Since the model is more flexible the posterior mean is free to shift towards the "true" average, being more "sensitive" to the data. This fact is more evident in cases B and C, where ε is random: the posterior mode of the number of clusters is around 10, while in case A is around 16. Here, the data determine the degree of approximation such that unreasonable a-priori information impacts the resulting number of groups less.

Furthermore, we fix $\sigma = 0.1$ and $\kappa = 0.45$ but we consider $\varepsilon \sim Gamma(\alpha, \beta)$, with support over all positive real numbers; in particular, we choose $(\alpha, \beta) \in \{(0.5, 2), (0.01, 0.1), (1, 10)\}$. The first combination corresponds to a relatively large mean (0.25) and variance (0.125) for ε. However, a large mass of the distribution lies around 0 due to the presence of an asymptote in the prior distribution. The second and third combinations have the same mean (0.1) but the variance is 1 and

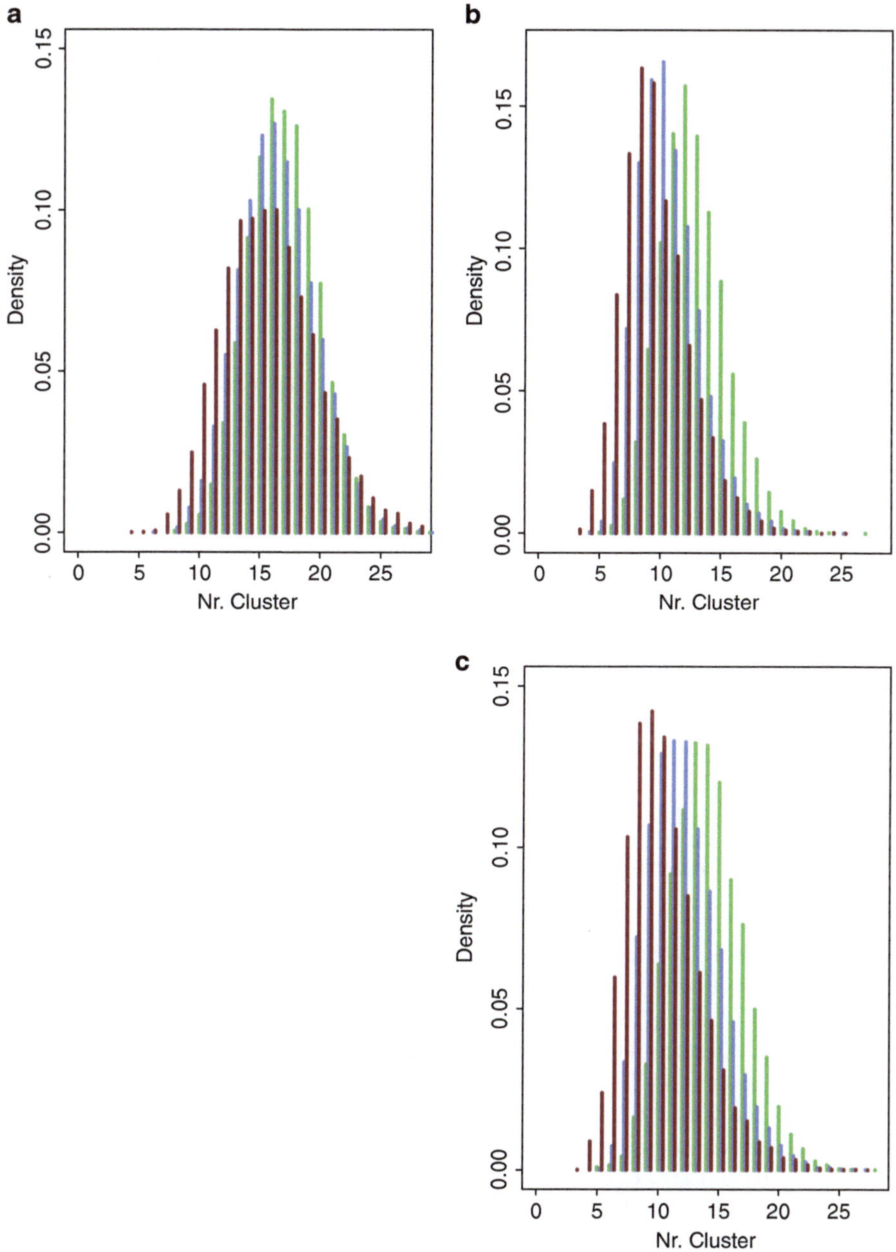

Fig. 2.3 Histograms of the posterior number of clusters in tests A, B, C. In *magenta* the tests with a bigger a-priori variance for K_n, in *green* the tests corresponding to a relatively small variance a-priori, in *blue* the intermediate ones. (**a**) $A7$, $A8$, $A9$. (**b**) $B7$, $B8$, $B9$. (**c**) $C7$, $C8$, $C9$

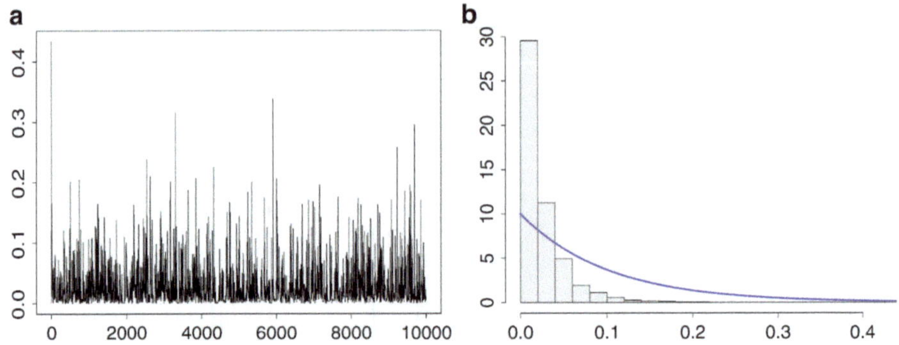

Fig. 2.4 (**a**) Traceplots and histogram for variable ε in the second test case. In panel (**b**) the *violet line* shows the prior distribution, i.e. $\varepsilon \sim gamma(1,10)$

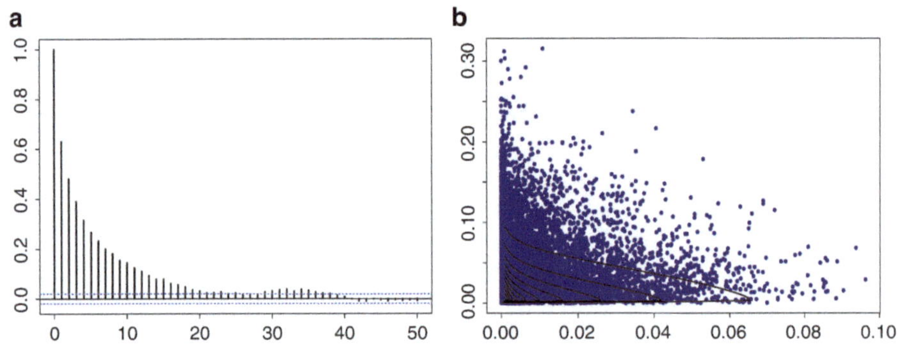

Fig. 2.5 Autocorrelation of variable ε, (**a**), and scatterplot of ε versus σ, (**b**): the *gray lines* represent the contour levels of the prior

0.01, respectively. We report for brevity only results for $(\alpha,\beta) = (1,10)$; however, we point out that some mixing problems in the chain for ε arise, when increasing the a-priori variance. Figure 2.4b shows that ε moves a posteriori towards smaller values with respect to the prior information. Besides, the traceplot of ε, Fig. 2.4a, exhibits a good mixing for the chain in this case.

Finally, we mention a further test, where all three parameters are random: in particular, we assume $\varepsilon \sim Beta(0.69, 2.06)$ with support on $(0, 0.1)$, $\sigma \sim Beta(1.1, 30)$ and $\kappa \sim Gamma(1.1, 8)$. The density estimate is satisfying, the only issue to mention is the high autocorrelation of ε and the correlation between the two parameters σ and ε (Fig. 2.5). This result is even more pronounced under a less informative prior distribution for (σ, ε).

2.5 Conclusions

A method to deal with a particularly flexible nonparametric mixture model, namely the NGG mixture model, is presented. It is based on a-priori truncation of the infinite sum defining the random probability measure P and it allows to computationally handle the presence of an infinite dimensional parameter, P, in the mixture model. In fact, conditionally on a threshold value ε, we can define a new process P_ε, which consists of a finite sum. We showed an application to density estimation for the popular Galaxy dataset. Through the exposition of several choices of the hyperparameters we established the robustness of the model and studied the relationship between posterior estimates and prior elicitation. In particular, we illustrated some suitable priors for the threshold parameter ε, letting in this case, the data drive the degree of approximation. If there is no need to consider a fully nonparametric model, ε may be relatively far from 0, implying smaller computational effort. Overall, density estimates were satisfying in all the experiments.

References

[1] Argiento, R., Guglielmi, A., Pievatolo, A.: Bayesian density estimation and model selection using nonparametric hierarchical mixtures. Comput. Stat. Data Anal. **54**(4), 816–832 (2010)

[2] Argiento, R., Bianchini, I., Guglielmi, A.: A blocked Gibbs sampler for NGG-mixture models via a-priori truncation. Stat. Comput. Advance online publication. doi: 10.1007/s11222-015-9549-6 (2015)

[3] Escobar, M.D., West, M.: Bayesian density estimation and inference using mixtures. J. Am. Stat. Assoc. **90**(430), 577–588 (1995)

[4] Griffin, J.E.: An adaptive truncation method for inference in bayesian nonparametric models. arXiv preprint arXiv:1308.2045 (2013). doi:10.1007/s11222-014-9519-4

[5] Griffin, J.E., Walker, S.G.: Posterior simulation of normalized random measure mixtures. J. Comput. Graph. Stat. **20**(1), 241–259 (2011)

[6] Gelfand, A.E., Kottas, A.: A computational approach for full nonparametric Bayesian inference under Dirichlet process mixture models. J. Comput. Graph. Stat. **11**(2), 289–305 (2002)

[7] Ishwaran, H., James, L.F.: Gibbs sampling methods for stick-breaking priors. J. Am. Stat. Assoc. **96**(453), 161–173 (2001)

[8] Kingman, J.F.C.: Poisson Processes. Oxford Studies in Probability, vol. 3. The Clarendon Press/Oxford University Press, New York (1993). Oxford Science Publications

[9] Lijoi, A., Mena, R.H., Prünster, I.: Controlling the reinforcement in Bayesian non-parametric mixture models. J. R. Stat. Soc. Ser. B Stat. Methodol. **69**(4), 715–740 (2007)

[10] MacEachern, S.N.: Computational methods for mixture of Dirichlet process models. In: Practical Nonparametric and Semiparametric Bayesian Statistics. Lecture Notes in Statistics, vol. 133, pp. 23–43. Springer, New York (1998)

[11] Müller, P., Mitra, R.: Bayesian nonparametric inference—why and how. Bayesian Anal. **8**(2), 269–302 (2013)

[12] Neal, R.M.: Markov chain sampling methods for Dirichlet process mixture models. J. Comput. Graph. Stat. **9**(2), 249–265 (2000)

[13] Papaspiliopoulos, O., Roberts, G.O.: Retrospective Markov chain Monte Carlo methods for Dirichlet process hierarchical models. Biometrika **95**(1), 169–186 (2008)
[14] Regazzini, E., Lijoi, A., Prünster, I.: Distributional results for means of normalized random measures with independent increments. Ann. Stat. **31**(2), 560–585 (2003). Dedicated to the memory of Herbert E. Robbins
[15] Walker, S.G.: Sampling the Dirichlet mixture model with slices. Commun. Stat. Simul. Comput. **36**(1–3), 45–54 (2007)

Chapter 3
Distributed Estimation of Mixture Models

Kamil Dedecius and Jan Reichl

Abstract The contribution deals with sequential distributed estimation of global parameters of normal mixture models, namely mixing probabilities and component means and covariances. The network of cooperating agents is represented by a directed or undirected graph, consisting of vertices taking observations, incorporating them into own statistical knowledge about the inferred parameters and sharing the observations and the posterior knowledge with other vertices. The aim to propose a computationally cheap online estimation algorithm naturally disqualifies the popular (sequential) Monte Carlo methods for the associated high computational burden, as well as the expectation-maximization (EM) algorithms for their difficulties with online settings requiring data batching or stochastic approximations. Instead, we proceed with the quasi-Bayesian approach, allowing sequential analytical incorporation of the (shared) observations into the normal inverse-Wishart conjugate priors. The posterior distributions are subsequently merged using the Kullback–Leibler optimal procedure.

Key words: Mixture estimation, Distributed estimation, Quasi-Bayesian estimation

3.1 Introduction

The rapid development of ad-hoc networks and the emergence of the so-called big data phenomenon have brought new challenges for distributed statistical data processing. For instance, the processing often needs to be decentralized, i.e. without any dedicated unit in the network. Instead, *all* agents are responsible for (i) taking measurements, (ii) processing them, and (iii) sharing the statistical knowledge about the (usually global) inferred parameters. In addition, the estimation should run

K. Dedecius (✉) • J. Reichl
Institute of Information Theory and Automation, Academy of Sciences of the Czech Republic, Prague, Czech Republic
e-mail: dedecius@utia.cas.cz; reichja3@fjfi.cvut.cz

S. Frühwirth-Schnatter et al. (eds.), *Bayesian Statistics from Methods to Models and Applications*, Springer Proceedings in Mathematics & Statistics 126,
DOI 10.1007/978-3-319-16238-6_3

online in many cases. This means to take observations of a dynamic process and incorporate them sequentially into the shared knowledge. This often disqualifies the popular sequential Monte Carlo (MC) approach due to the associated high computational burden. Excellent surveys on distributed estimation are the recent papers by Sayed [10] (non-MC) and Hlinka et al. [5] (MC-based).

Despite the great potential of the Bayesian paradigm in this field, its adoption is still rather the exception than the rule. From the probabilistic viewpoint, the resulting "classical" (that is, non-Bayesian) algorithms often suffer from statistical inconsistencies. For instance, point estimators are often combined without reflecting the associated uncertainty, which may lead to erroneous estimates. The first author's work [1] aims at partially filling this gap. It proposes a fully Bayesian approach to decentralized distributed estimation with fusion, based on minimizing the Kullback–Leibler divergence. The present contribution extends the results to the case of mixture models, covered for the static cases, e.g., in [3, 8, 13].

The novelty of the proposed framework lies in a fully analytical Bayesian processing of observations and shared knowledge about the estimated parameters. To this end, the underlying theory relies on the quasi-Bayesian approach, proposed by Smith, Makov, and Titterington [11, 12] and followed by Kárný et al. [6], whose approach is adopted here. It provides analytical tractability of mixture inference by relying on point estimators where necessary. Though we focus on normal mixtures, the results are applicable to homogeneous mixtures of exponential family distributions.

3.2 Quasi-Bayesian Estimation of Mixture Models

Consider an observable time series $\{Y_t,\ t \in \mathbb{N}\}$ with $Y_t \in \mathbb{R}^n$ following a normal mixture distribution

$$\begin{aligned} Y_t|\phi,\theta &\sim \phi_1 \mathrm{N}(\mu_1,\Sigma_1) + \ldots + \phi_K \mathrm{N}(\mu_K,\Sigma_K) \\ &\sim \phi_1 \mathrm{N}(\theta_1) + \ldots + \phi_K \mathrm{N}(\theta_K), \end{aligned} \tag{3.1}$$

where $\mathrm{N}(\mu_k,\Sigma_k)$ denotes the kth component density, namely a normal distribution with mean vector $\mu_k \in \mathbb{R}^n$ and covariance matrix $\Sigma_k \in \mathbb{R}^{n\times n}$, in the latter notation summarized by $\theta_k = \{\mu_k,\Sigma_k\}$. The nonnegative variables ϕ_k taking values in the unit K-simplex are the component probabilities. The number of components K is assumed to be known a priori. Furthermore, the notation $\theta = \{\theta_1,\ldots,\theta_K\}$, $\phi = \{\phi_1,\ldots,\phi_K\}$ is used.

Let $p_k(y_k|\theta_k)$ be the probability density function of the kth component, yielding the mixture density of the form

$$p(y_t|\theta,\phi) = \sum_{k=1}^{K} \phi_k p_k(y_t|\theta_k). \tag{3.2}$$

At each time instant t the observation y_t is generated by the k_tth component $p_k(y_t|\theta_k)$, selected with probability ϕ_k,

$$p(y_t|\theta,\phi,k_t) = \prod_{k=1}^{K} [\phi_k p_k(y_t|\mu_k,\Sigma_k)]^{S_{k,t}}, \tag{3.3}$$

where $S_{k,t}$ is the indicator function of the active component

$$S_{k,t} = \begin{cases} 1 & \text{if } S_{k,t} = k_t, \\ 0 & \text{otherwise.} \end{cases} \tag{3.4}$$

In other words, $S_t = (S_{1,t},\ldots,S_{K,t})$ can be viewed as a vector with 1 on the k_tth position and zeros elsewhere, and hence follows the multinomial distribution Multi$(1,\phi)$.

From the Bayesian viewpoint the topological property of ϕ is crucial, as it allows its modelling with the Dirichlet distribution with parameters $\kappa_1,\ldots,\kappa_K$,

$$\phi = (\phi_1,\ldots,\phi_K) \sim \text{Dir}(\kappa_1,\ldots,\kappa_K), \qquad \kappa_k > 0 \quad \text{for all} \quad k = 1,\ldots,K,$$

conjugate to the multinomial distribution of S_t. Sequential estimation of each single component mean and covariance can then proceed with the conjugate normal inverse-Wishart distribution (or normal inverse-gamma in the univariate case),

$$\theta_k = \{\mu_k,\Sigma_k\} \sim \text{NiW}(m,s,a,b), \qquad m \in \mathbb{R}^n,\ \ s \in \mathbb{R}^{n\times n},\ \ a,b > 0.$$

Exact knowledge of S_t would make the Bayesian inference of both the component parameters μ_k,Σ_k and mixing probabilities ϕ easily tractable, since the product (3.3) simplifies to a single density and a single component probability. Likewise, the Bayesian inference of mixing probabilities ϕ is easy under known components, as the detection of the active one is a relatively simple hypotheses testing problem, see, e.g., [4]. However, our attention is shifted towards estimating both component parameters μ,Σ and mixing probabilities ϕ. For this sake, we need to derive the Bayesian update

$$\pi_{\phi,\theta}(\phi,\theta|y_{1:t},k_{1:t}) \propto \pi_{\phi,\theta}(\phi,\theta|y_{1:t-1},k_{1:t-1}) \prod_{k=1}^{K} [\phi_k p_k(y_t|\theta_k)]^{S_{k,t}}$$

where the joint prior distribution is assumed to be

$$\pi_{\phi,\theta}(\phi,\theta|y_{1:t-1},k_{1:t-1}) = \pi_\phi(\phi|y_{1:t-1},k_{1:t-1})\pi_\theta(\theta|y_{1:t-1},k_{1:t-1}).$$

The independence of ϕ and θ allows tractable computation of the posterior distribution. Indeed, this assumption is not quasi-Bayes specific.

In this case, Kárný et al. [6] propose to rely on the approach of Smith, Makov, and Titterington [11, 12] and replace the latent indicators $S_{k,t}$ defined in Eq. (3.4) by their respective point estimates with respect to ϕ_k and θ_k of the form

$$\begin{aligned} \widehat{S}_{k,t} &= \mathbb{E}\left[S_{k,t}|y_{1:t}, k_{1:t-1}\right] \\ &\propto \mathbb{E}\left[\phi_k|y_{1:t-1}, k_{1:t-1}\right] p_k(y_t|y_{1:t-1}, k_{1:t-1}), \end{aligned} \tag{3.5}$$

where

$$p_k(y_t|y_{1:t-1}, k_{1:t-1}) = \int p_k(y_t|\theta_k)\pi_{\theta_k}(\theta_k|y_{1:t-1}, k_{1:t-1})\mathrm{d}\theta_k \tag{3.6}$$

is the predictive distribution (under normal inverse-Wishart prior it is a Student's t distribution). To summarize, the estimation of the indicator $S_{k,t}$ of the active component k is based on (i) testing the component membership based on the predictive likelihood (3.6), and (ii) the estimated probability of the particular component $\mathbb{E}[\phi_k|\cdot]$ in (3.5).

The quasi-Bayesian update then takes the weighted form of the regular update under known S_t,

$$\pi_\phi(\phi|y_{1:t}, k_{1:t}) \propto \mathbb{E}\left[\widehat{S}_t|y_{1:t}, k_{1:t-1}\right] \pi_\phi(\phi|y_{1:t-1}, k_{1:t-1}), \tag{3.7}$$

$$\pi_{\theta_k}(\theta_k|y_{1:t}, k_{1:t}) \propto \left[p_k(y_t|\theta_k)\right]^{\widehat{S}_{k,t}} \pi_{\theta_k}(\theta_k|y_{1:t-1}, k_{1:t-1}). \tag{3.8}$$

If the component density is rewritten as the exponential family and the prior density is conjugate, then, as shown in the Appendix, the update of the relevant hyperparameters is particularly easy.

3.3 Distributed Estimation

Assume that the distributed estimation runs in a network represented by a directed or undirected connected graph $\mathrm{G}(V,E)$ consisting of a set of vertices $V = \{1,\ldots,N\}$ (also called nodes or agents) and a set E of edges, defining the graph topology. The vertices $n \in V$ are allowed to communicate with adjacent vertices. For a fixed vertex n, these neighbors form a complete bipartite subgraph (every neighboring vertex is connected with n) with radius 1, diameter at most 2 and of type star (unless the vertex n is of degree 1), where n is the central vertex and all other vertices peripheral. The set of vertices of this subgraph is denoted by V_n.

The vertices independently observe the process $\{Y_t, t \in \mathbb{N}\}$, taking observations $y_t^{(n)}, n \in V$. These are shared within V_n in the sense that each vertex n has access to $y_t^{(j)}$ of vertices $j \in V_n$ and incorporates them according to the quasi-Bayesian estimation theory outlined in the previous section. That is, each node n ends with the joint posterior density

$$\pi_{\phi,\theta}^{(n)}(\phi,\theta|\widetilde{y}_{1:t},\widetilde{k}_{1:t}), \tag{3.9}$$

resulting from the number of $\text{card}(V_n)$ updates of the form (3.7) and (3.8). Here, tilde denotes the statistical knowledge comprising the V_n's information relevant to the particular variable. This step is called *adaptation*, e.g., [10].[1]

3.3.1 Combination of Estimates

In the *combination* step [10], the vertices $n \in V$ access V_n's posterior distributions (3.9) resulting from the adaptation,

$$\pi_{\phi,\theta}^{(j)}(\phi,\theta|\widetilde{y}_{1:t},\widetilde{k}_{1:t}), \qquad j \in V_n.$$

Now the goal is to represent (i.e., approximate) them by a *single* joint posterior $\widetilde{\pi}_{\phi,\theta}^{(n)}$ parameterizing the mixture (3.1) in consideration. To this end, we adopt the Kullback–Leibler divergence [7] defined in the Appendix, and seek for $\widetilde{\pi}_{\phi,\theta}^{(n)}$ satisfying

$$\sum_{j\in V_n} \alpha_{nj} \mathrm{D}(\widetilde{\pi}_{\phi,\theta}^{(n)}||\pi_{\phi,\theta}^{(j)}) \to \min, \tag{3.10}$$

where $\alpha_{nj} = 1/(\text{card}(V_n))$ are nonnegative uniform weights assigned to nodes $j \in V_n$ summing to unity. Other weight choices, e.g. reflecting properties of the neighboring vertices are possible as well.

Let us impose an additional assumption simplifying the theory: identical order of component parameters and significantly overlapping densities $\pi_{\phi,\theta}^{(j)}$ of all $j \in V_n$. This means that the order of the components and their parameterization agrees at all vertices in V_n (and hence V). This assumption can be easily removed by incorporating detection of similar posterior distributions or enforced by starting from identical initial priors.

We exploit the following general proposition proved, e.g., in [1]. Although we consider exponential family distributions (where it provides analytically tractable results), the proposition is not limited to them.

Proposition 1. *Let* $\pi_{\phi,\theta}^{(j)}$ *be the posterior probability density functions of vertices* $j \in V_n$ *and* α_{nj} *their weights from the unit* $\text{card}(V_n)$*-simplex. Their approximation by a single density* $\widetilde{\pi}_{\phi,\theta}^{(n)}$ *optimal in the Kullback–Leibler sense* (3.10) *has the form*

---

[1] The terms "adaptation" and "combination" were introduced by [10]. We adopt them for our Bayesian counterparts.

$$\widetilde{\pi}_{\phi,\theta} \propto \prod_{j\in V_n} \left[\pi_{\phi,\theta}^{(j)}\right]^{\alpha_{nj}}. \tag{3.11}$$

The resulting approximate posterior density hence virtually parameterizes a much richer mixture, however, the individual densities overlap by the given assumption. Then Proposition 1 gives a method for reduction to the parametrization of K components,

$$\widetilde{\pi}_{\phi}^{(n)} \propto \prod_{j\in V_n} \left[\pi_{\phi}^{(j)}\right]^{\alpha_{nj}} \qquad \text{and} \qquad \widetilde{\theta}_{\phi}^{(n)} \propto \prod_{j\in V_n} \left[\theta_{\phi}^{(j)}\right]^{\alpha_{nj}},$$

which, due to the structure of the conjugate priors (see Appendix) and component ordering yields

$$\widetilde{\xi}_{k,t}^{(n)} = \sum_{j\in V_n} \alpha_{nj}\xi_{k,t}^{(j)}, \quad \widetilde{\nu}_{k,t}^{(n)} = \sum_{j\in V_n} \alpha_{nj}\nu_{k,t}^{(j)}, \qquad \text{and} \qquad \widetilde{\kappa}_{k,t}^{(n)} = \sum_{j\in V_n} \alpha_{nj}\kappa_{k,t}^{(j)},$$

for the hyperparameters ξ, ν and κ of the prior distributions for θ and ϕ, respectively. The resulting KL-optimal posterior is then again conjugate to the model and can be used for the subsequent adaptation step.

3.4 Simulation Example

The simulation example deals with estimating a three-component normal mixture model, for simplicity univariate of the form

$$Y \sim \frac{1}{3}\mathrm{N}(-2,1) + \frac{1}{3}\mathrm{N}(4,1) + \frac{1}{3}\mathrm{N}(8,2),$$

with unknown means and variances. The graph $\mathrm{G}(V,E)$, whose scheme is depicted together with the components and samples in Fig. 3.1, consists of a set of vertices $V = \{1,\ldots,8\} \setminus \{6\}$. The sixth vertex is disconnected and serves for comparison. The vertices $n \in V \cup \{6\}$ take observations $y_t^{(n)}$ with $t = 1,\ldots,300$. Clearly, one would expect relatively easy identification of the leftmost component, while the other two may be problematic due to their closeness. The quasi-Bayesian estimation of components $k \in \{1,2,3\}$ exploits the conjugate normal inverse-gamma prior $\mathrm{NIG}(\mu_k,\sigma_k;m_k,s_k,a_k,b_k) = \mathrm{N}(\mu_k|\sigma_k^2;m_k,\sigma^2 s_k) \times \mathrm{IG}(\sigma_k^2;a_k,b_k)$ with initial hyperparameters m_k set to 0, 3, and 6, respectively; the other hyperparameters are $s_k = 1, a_k = 2, b_k = 2$ for all k. The prior for the component probabilities is $\phi \sim \mathrm{Dir}(\frac{1}{3},\frac{1}{3},\frac{1}{3})$. This initialization is identical across the graph.

The progress of the point estimates of μ_k and σ_k is depicted in Fig. 3.2 for the isolated vertex 6 (left) and the randomly chosen vertex 4 (right). The point estimates of μ_k converge relatively well in both cases, however, the variance estimates

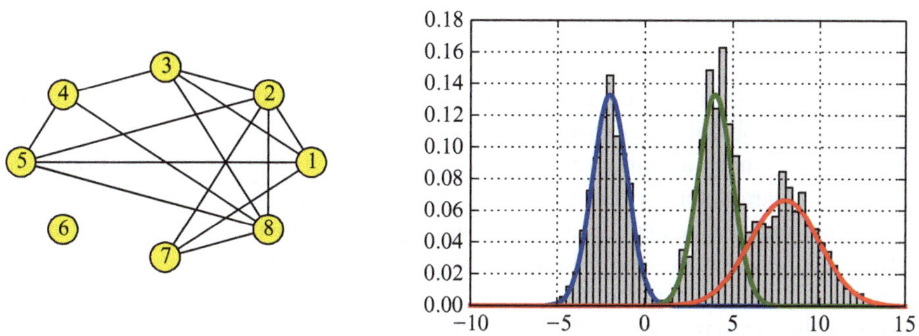

Fig. 3.1 *Left*: Layout of the graph with isolated node 6 for comparison. *Right*: Normalized histogram and true components of the mixture

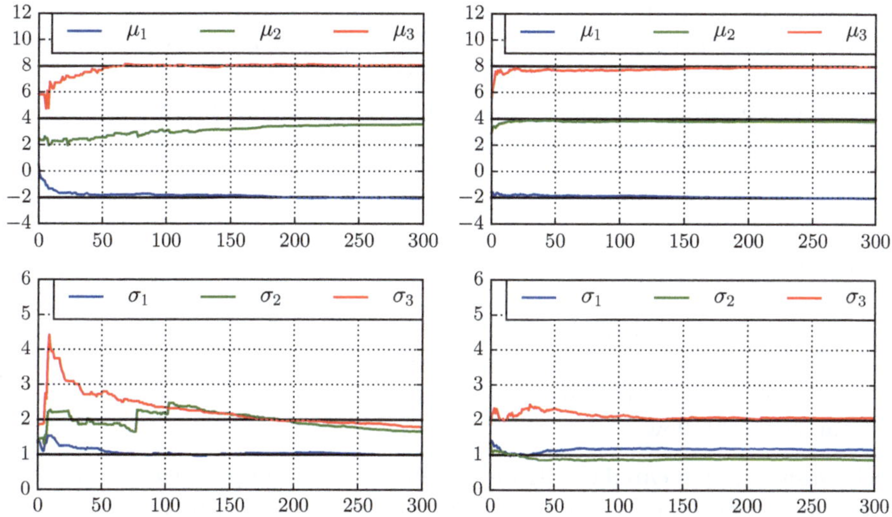

Fig. 3.2 Evolution of estimates of component means and standard deviations. *Left*: isolated vertex 6. *Right*: situation at a chosen cooperating vertex 4. *Solid black lines* depict true values

Table 3.1 Statistics of mean square errors (MSEs) of resulting estimates: distributed estimation and isolated vertex 6

MSE	Min (distr.)	Max (distr.)	Mean (distr.)	Vertex 6
Means μ_k	0.007	0.007	0.007	0.057
Variances σ_k^2	0.092	0.481	0.222	1.26
Comp. probabilities ϕ	0	0	0	0.001

converge well only in the case of the distributed estimation (with the exception of σ_1^2). This is due to the much richer data available for the interconnected vertices. The mean squared errors (MSE) of the final estimates are given in Table 3.1.

3.5 Conclusion

The quasi-Bayesian method for analytically tractable sequential inference of parameters of probabilistic mixtures has been extended to the case of distributed estimation of normal mixture model with unknown mixing probabilities and component parameters. Here, *distributed* means that there is a graph (network) of cooperating vertices (nodes, agents) sharing their statistical knowledge (observations and estimates) with a limited subset of other vertices. This knowledge is combined at each vertex: the observations are incorporated by means of the Bayes' theorem, the estimates are combined via the Kullback–Leibler optimal rule.

The main advantage of the method is its simplicity and scalability. Unlike Monte Carlo approaches, it is computationally very cheap. The authors have recently shown in [2] that this method is suitable for the whole class of mixture models consisting of exponential family distributions and their conjugate prior distributions.

One difficulty associated with the method is common for most mixture estimation methods, namely the initialization. In addition, merging and splitting of components after the combination of estimates would significantly enhance the suitability of the approach for dynamic cases. These topics remain for further research.

Appendix

Below we give several useful definitions and lemmas regarding the Bayesian estimation of exponential family distributions with conjugate priors [9]. The proofs are trivial. Their application to the normal model and normal inverse-gamma prior used in Sect. 3.4 follows.

Definition 1 (Exponential family distributions and conjugate priors). Any distribution of a random variable y parameterized by θ with the probability density function of the form

$$p(y|\theta) = f(y)g(\theta)\exp\{\eta(\theta)^{\intercal}T(y)\},$$

where f, g, η, and T are known functions, is called an exponential family distribution. $\eta \equiv \eta(\theta)$ is its natural parameter, $T(y)$ is the (dimension preserving) sufficient statistic. The form is not unique.

Any prior distribution for θ is said to be conjugate to $p(y|\theta)$, if it can be written in the form

$$\pi(\theta|\xi,\nu) = q(\xi,\nu)g(\theta)^{\nu}\exp\{\eta(\theta)^{\intercal}\xi\},$$

where q is a known function and the hyperparameters $\nu \in \mathbb{R}^{+}$ and ξ is of the same shape as $T(y)$.

Lemma 1 (Bayesian update with conjugate priors). *Bayes' theorem*

$$\pi(\theta|\xi_t, \nu_t) \propto p(y_t|\theta)\pi(\theta|\xi_{t-1}, \nu_{t-1})$$

yields the posterior hyperparameters as follows:

$$\xi_t = \xi_{t-1} + T(y_t) \qquad \text{and} \qquad \nu_t = \nu_{t-1} + 1.$$

Lemma 2. *The normal model*

$$p(y_t|\mu,\sigma^2) = \frac{(\sigma^2)^{-\frac{1}{2}}}{\sqrt{2\pi}} \exp\left\{-\frac{1}{2\sigma^2}(y_t - \mu)^2\right\}$$

where μ, σ^2 are unknown can be written in the exponential family form with

$$\eta = \left(\frac{\mu}{\sigma^2}, \frac{-1}{2\sigma^2}, \frac{-\mu^2}{2\sigma^2}\right)^\intercal, \qquad T(y_t) = (y, y^2, 1)^\intercal, \qquad g(\eta) = (\sigma^2)^{-\frac{1}{2}}.$$

Lemma 3. *The normal inverse-gamma prior distribution for μ, σ^2 with the (non-natural) real scalar hyperparameters m, and positive s, a, b, having the density*

$$p(\mu, \sigma^2|m, s, a, b) = \frac{b^a(\sigma^2)^{a+1+\frac{1}{2}}}{\sqrt{2\pi s}\Gamma(a)} \exp\left\{-\frac{1}{\sigma^2}\left[b + \frac{1}{2s}(m-\mu)^2\right]\right\}$$

can be written in the prior-conjugate form with

$$\xi_t = \left(\frac{m}{s}, \frac{m^2}{s} + 2b, \frac{1}{s}\right)^\intercal.$$

Lemma 4. *The Bayesian update of the normal inverse-gamma prior following the previous lemma coincides with the 'ordinary' well-known update of the original hyperparameters,*

$$s_t^{-1} = s_{t-1}^{-1} + 1, \qquad a_t = a_{t-1} + \frac{1}{2},$$

$$m_t = s_t\left(\frac{m_{t-1}}{s_{t-1}} + y_t\right), \qquad b_t = b_{t-1} + \frac{1}{2}\left(\frac{m_{t-1}^2}{s_{t-1}} - \frac{m_t^2}{s_t} + y_t^2\right).$$

Definition 2 (Kullback–Leibler divergence). Let $f(x), g(x)$ be two probability density functions of a random variable x, f absolutely continuous with respect to g. The Kullback–Leibler divergence is the nonnegative functional

$$D(f||g) = \mathbb{E}_f \left[\log \frac{f(x)}{g(x)} \right] = \int f(x) \log \frac{f(x)}{g(x)} dx, \tag{3.12}$$

where the integration domain is the support of f. The Kullback–Leibler divergence is a premetric; it is zero if $f = g$ almost everywhere, it does not satisfy the triangle inequality nor is it symmetric.

Acknowledgements This work was supported by the Czech Science Foundation, postdoctoral grant no. 14–06678P. The authors thank the referees for their valuable comments.

References

[1] Dedecius, K., Sečkárová, V.: Dynamic diffusion estimation in exponential family models. IEEE Signal Process. Lett. **20**(11), 1114–1117 (2013)
[2] Dedecius, K., Reichl, J., Djurić, P.M.: Sequential estimation of mixtures in diffusion networks. IEEE Signal Process. Lett. **22**(2), 197–201 (2015)
[3] Dongbing, Gu.: Distributed EM algorithm for Gaussian mixtures in sensor networks. IEEE Trans. Neural Netw. **19**(7), 1154–1166 (2008)
[4] Frühwirth-Schnatter, S.: Finite Mixture and Markov Switching Models. Springer, London (2006)
[5] Hlinka, O., Hlawatsch, F., Djurić, P.M.: Distributed particle filtering in agent networks: a survey, classification, and comparison. IEEE Signal Process. Mag. **30**(1), 61–81 (2013)
[6] Kárný, M., Böhm, J., Guy, T.V., Jirsa, L., Nagy, I., Nedoma, P., Tesař, L.: Optimized Bayesian Dynamic Advising: Theory and Algorithms. Springer, London (2006)
[7] Kullback, S., Leibler, R.A.: On information and sufficiency. Ann. Math. Stat. **22**(1), 79–86 (1951)
[8] Pereira, S.S., Lopez-Valcarce, R., Pages-Zamora, A.: A diffusion-based EM algorithm for distributed estimation in unreliable sensor networks. IEEE Signal Process. Lett. **20**(6), 595–598 (2013)
[9] Raiffa, H., Schlaifer, R.: Applied Statistical Decision Theory (Harvard Business School Publications). Harvard University Press, Cambridge (1961)
[10] Sayed, A.H.: Adaptive networks. Proc. IEEE **102**(4), 460–497 (2014)
[11] Smith, A.F.M., Makov, U.E.: A Quasi-Bayes sequential procedure for mixtures. J. R. Stat. Soc. Ser. B (Methodol.) **40**(1), 106–112 (1978)
[12] Titterington, D.M., Smith, A.F.M., Makov, U.E.: Statistical Analysis of Finite Mixture Distributions. Wiley, New York (1985)
[13] Weng, Y., Xiao, W., Xie, L.: Diffusion-based EM algorithm for distributed estimation of Gaussian mixtures in wireless sensor networks. Sensors **11**(6), 6297–316 (2011)

Chapter 4
Jeffreys' Priors for Mixture Estimation

Clara Grazian and Christian P. Robert

Abstract Mixture models may be a useful and flexible tool to describe data with a complicated structure, for instance characterized by multimodality or asymmetry. The literature about Bayesian analysis of mixture models is huge, nevertheless an "objective" Bayesian approach for these models is not widespread, because it is a well-established fact that one needs to be careful in using improper prior distributions, since the posterior distribution may not be proper, yet noninformative priors are often improper. In this work, a preliminary analysis based on the use of a dependent Jeffreys' prior in the setting of mixture models will be presented. The Jeffreys' prior which assumes the parameters of a Gaussian mixture model is shown to be improper and the conditional Jeffreys' prior for each group of parameters is studied. The Jeffreys' prior for the complete set of parameters is then used to approximate the derived posterior distribution via a Metropolis–Hastings algorithm and the behavior of the simulated chains is investigated to reach evidence in favor of the properness of the posterior distribution.

Key words: Improper priors, Mixture of distributions, Monte Carlo methods, Noninformative priors

---

C. Grazian (✉)
CEREMADE, Université Paris-Dauphine, Paris, France

Dipartimento di Scienze Statistiche, Sapienza Università di Roma, Rome, Italy
e-mail: grazian@ceremade.dauphine.fr

C.P. Robert
CEREMADE, Université Paris-Dauphine, Paris, France

Department of Statistics, University of Warwick, Coventry, UK
e-mail: xian@ceremade.dauphine.fr

S. Frühwirth-Schnatter et al. (eds.), *Bayesian Statistics from Methods to Models and Applications*, Springer Proceedings in Mathematics & Statistics 126,
DOI 10.1007/978-3-319-16238-6_4

4.1 Introduction

The probability density function of the random variable $\mathbf{x}$ of a mixture model is given as follows:

$$g(\mathbf{x} \mid \psi) = \sum_{i=1}^{K} w_i f_i(\mathbf{x} \mid \theta_i), \tag{4.1}$$

where $\mathbf{x}$ is a random variable with probability density function $g(\cdot)$ which depends on a parameter vector $\psi = (\theta_1, \cdots, \theta_K, w_1, \cdots, w_K)$, where $w_i \in (0,1)$ and $\sum_{i=1}^{K} w_i = 1$, K is the number of components and θ_i is the vector of parameters of the ith component, whose behavior is described by the density function $f_i(\cdot)$.

In this setting, the maximum likelihood estimation may be problematic, even in the simple case of Gaussian mixture models, as shown in [2]. For a comprehensive review, see [11]. In a Bayesian setting, [4] and [6] suggest to be careful when using improper priors, in particular because it is always possible that the sample does not include observations for one or more components, thus the data are not informative about those particular components. Avoiding improper priors is not necessary, however the properness of the posterior distribution has to be proven and works exist which show that independent improper priors on the parameters of a mixture model lead to improper posteriors, except for the case where one component has no observation in the sample is prohibited (as in [3]). Some proposals of "objective priors" in the setting of mixture models are the partially proper priors in [3, 7] and [10], the data dependent prior in [12] and the weakly informative prior in [8], which may or may not be data-dependent.

In this work, we want to analyze the posterior distribution for the parameters of a mixture model with a finite number of components when the Jeffreys' definition of a noninformative prior (see [5]) is applied. In particular, we want to assess the convergence of the Markov chain derived from an MCMC approximation of the posterior distribution when using the Jeffreys' prior for the parameters of a Gaussian mixture model, even when the prior for some parameters is improper conditional on the others.

The outline of the paper is as follows: in Sect. 4.2, the Jeffreys' prior is presented and an explanation about the reason why improper priors have to be used with care in the setting of mixture models is given; then, the Jeffreys' prior for the weights of a general mixture model conditional on the other parameters is presented in Sect. 4.2.1 and the Jeffreys' priors for the means and the standard deviations when every other parameter is known are presented in Sect. 4.2.2. Section 4.3 describes the algorithms used to implement simulations. Section 4.4 shows the results for the posterior distributions obtained when using a dependent Jeffreys' prior for all the parameters of a Gaussian mixture model based on simulations, in particular including an example for a three-component Gaussian mixture model; finally, Sect. 4.5 concludes the paper with a discussion.

4.2 Jeffreys' Priors for Mixture Models

We recall that Jeffreys' prior was introduced by [5] as a default prior based on the Fisher information matrix $I(\theta)$ as

$$\pi^J(\theta) \propto |I(\theta)|^{\frac{1}{2}}, \tag{4.2}$$

whenever the latter is well defined. In most settings, Jeffreys' priors are improper, which may explain their conspicuous absence in the domain of mixture estimations, since the latter prohibits the use of most improper priors by allowing any subset of components to go empty. That is, the likelihood of a mixture model can always be decomposed into a sum over all possible partitions of the data with K groups at most, where K is the number of components of the mixture. This means that there are terms in this sum where no observation from the sample carries information about the parameters of a specific component. In particular, consider independent improper priors

$$\pi(\theta_1, \cdots, \theta_K) \propto \prod_{j=1}^{K} \pi^*(\theta_j), \tag{4.3}$$

such that $\int \pi^*(\theta_k) d\theta_k = \infty \ \forall \ k \in \{1, \cdots, K\}$. Mixture models are an example of latent variable models, where the density function may be rewritten in an augmented version as

$$g(\mathbf{x}; \psi) = \sum_{S \in \mathscr{S}_K} f_j(\mathbf{x}; S, \theta_j) \prod_{j=1}^{K} \pi^*(\theta_j) \pi(S \mid \mathbf{w}) \pi(\mathbf{w}), \tag{4.4}$$

where the summation runs over the set $\mathscr{S}_K$ of all the K^N possible classifications S. Then, if there is an empty component (let's say the jth), i.e. a component with no observation in the sample, the complete-data likelihood does not carry information about that particular component and the posterior distribution for it will depend only on the prior and will have an infinite integral, if the prior is improper:

$$\int \prod_{i:S_i=j} g(x_i; \theta_j) \pi^*(\theta_j) d\theta_j \propto \int \pi^*(\theta_j) d\theta_j = \infty. \tag{4.5}$$

Another obvious reason for the absence of Jeffreys' priors is a computational one, namely the closed-form derivation of the Fisher information matrix is almost inevitably impossible. The reason are integrals which cannot be analytically computed having the form

$$-\int_{\mathscr{X}} \frac{\partial^2 \log \left[\sum_{k=1}^{K} w_k f(\mathbf{x}|\theta_k)\right]}{\partial \theta_i \partial \theta_j} \left[\sum_{k=1}^{K} w_k f(\mathbf{x}|\theta_k)\right] \mathrm{d}x. \tag{4.6}$$

4.2.1 Jeffreys' Prior for the Weights of a Mixture Model

Consider a two-component mixture model with known parameters of the component distributions. The Jeffreys' prior for the weights is just a function of only one parameter because of the constraint on the sum of the weights:

$$\pi^J(w_1) \propto \sqrt{\int_{\mathscr{X}} \frac{(f(\mathbf{x};\theta_1) - f(\mathbf{x};\theta_2))^2}{w_1 f(\mathbf{x};\theta_1) + w_2 f(\mathbf{x};\theta_2)} d\mathbf{x}} \tag{4.7}$$

$$\leq \sqrt{\int_{\mathscr{X}} \frac{(f(\mathbf{x};\theta_1) - f(\mathbf{x};\theta_2))^2}{w_1 f(\mathbf{x};\theta_1)} d\mathbf{x}} \tag{4.8}$$

$$= \sqrt{\frac{1}{w_1} c_1}, \tag{4.9}$$

where c_1 is a positive constant and $\mathscr{X}$ is the support of the random variable $\mathbf{x}$ which is modelled as a mixture. The resulting prior may be easily generalized to the case of K components for which the generic element of the Fisher information matrix is

$$\int_{\mathscr{X}} \frac{(f(\mathbf{x};\theta_i) - f(\mathbf{x};\theta_K))(f(\mathbf{x};\theta_j) - f(\mathbf{x};\theta_K))}{\sum_{k=1}^K w_k f(\mathbf{x};\theta_k)} d\mathbf{x}, \tag{4.10}$$

where $i \in \{1, \cdots, K-1\}$ and $j \in \{1, \cdots, K-1\}$. As shown above, this prior is proper and it is easy to see that it is convex by studying its second derivative (in the general case of K components, it can be shown that the prior is still proper because all the integrals in the Fisher information matrix are finite and the marginals are still convex). The form of the prior depends on the type of components. For an approximation of the prior and of the derived posterior distribution based on a sample of 100 observations for a particular choice of parameters, see Fig. 4.1. We have compared this approximation to the ones obtained by fixing the parameters at different values: the Jeffreys' prior for the weights of a mixture model is more symmetric as the components are more similar in terms of variance and it is more concentrated around the extreme values of the support as the means are more distant.

The Jeffreys' prior for the weights of a mixture model could be approximated by a beta distribution, which represents the traditional marginal distribution for the weights of a mixture model (since in the literature the Dirichlet distribution is, in general, the default choice as proper prior distribution): after having obtained a sample via Metropolis–Hastings algorithm which approximates the Jeffreys' prior (which is not known in closed form), the parameters of the approximating beta distribution may be estimated from the sample with the method of moments. Figure 4.2 shows this approximation for the weight of the first component of a Gaussian mixture model for an increasing number of components, each having the same standard deviation. It is evident from the figure that the (marginal) Jeffreys' distribution and its beta approximation tend to be more and more concentrated

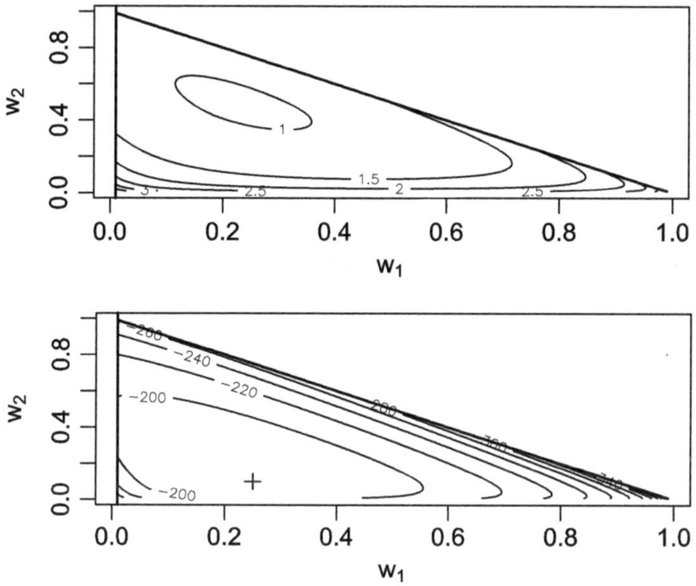

Fig. 4.1 Approximations of the (conditional) prior (*top*) and derived posterior (*bottom*) distributions for the weights of the three-component Gaussian mixture model $0.25 \cdot N(-1,1) + 0.10 \cdot N(0,5) + 0.65 \cdot N(2,0.5)$

Fig. 4.2 Beta approximations of the (conditional) prior distributions for the weight of the first component of a Gaussian mixture model with an increasing number K of components (with a fixed standard deviation equal to 1 for all the components and location parameters chosen as the first K elements of $\{-9,9,0,-6,6,-3,3\}$)

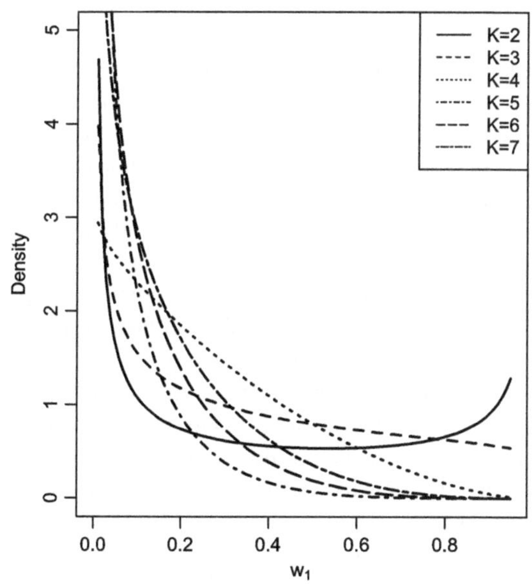

around 0 as the number of components increases. Both the variance and the mean of the beta approximations tend to stabilize around values close to 0, while it is not evident that there is a particular behavior for the parameters of the beta distribution, which could be smaller or greater than 1.

4.2.2 *Jeffreys' Prior for the Means and the Standard Deviations of a Gaussian Mixture Model*

Consider a two-component Gaussian mixture model. The conditional Jeffreys' prior for the mean parameters depends on the following derivatives:

$$\frac{\partial^2 \log f}{\partial \mu_i^2} = \left\{ \frac{w_i N(\mu_i, \sigma_i) \left[\left(\frac{x-\mu_i}{\sigma_i^2} \right)^2 - \frac{1}{\sigma_i^2} \right]}{w_1 N(\mu_1, \sigma_1) + w_2 N(\mu_2, \sigma_2)} \right\} - \left\{ \frac{w_i N(\mu_i, \sigma_i) \exp\left(-\frac{1}{2} \left(\frac{x-\mu_i}{\sigma_i} \right) \right) \frac{x-\mu_i}{\sigma_i^2}}{w_1 N(\mu_1, \sigma_1) + w_2 N(\mu_2, \sigma_2)} \right\}^2, \tag{4.11}$$

for $i \in \{1,2\}$ and

$$\frac{\partial^2 \log f}{\partial \mu_1 \partial \mu_2} = - \frac{w_1 N(\mu_1, \sigma_1) \frac{x-\mu_1}{\sigma_1^2} \cdot w_2 N(\mu_2, \sigma_2) \frac{x-\mu_2}{\sigma_2^2}}{w_1 N(\mu_1, \sigma_1) + w_2 N(\mu_2, \sigma_2)}. \tag{4.12}$$

With a simple change of variable $y = x - \mu_i$ for some choice of $i \in \{1,2\}$, it is easy to see that each element of the Fisher information matrix depends only on the difference between the means but not on μ_i alone, therefore it is flat with respect to each μ_i. The generalization to K components is obvious.

An approximation of the prior and derived posterior distribution based on a sample of 100 observations of the means of a two-component Gaussian mixture model is shown in Fig. 4.3. When only the means are unknown, it is evident that the prior is constant on the difference between the means and is increasing with the difference of the means. On the contrary, the posterior distribution is concentrated around the true values and shows classical label switching. Again, Fig. 4.3 has been compared to similar approximations obtained by fixing the parameters at different values: the conditional Jeffreys' prior for the means of a Gaussian mixture model is more symmetric, as the standard deviations become more similar and the choice of the weights seems to influence only the approximation of the posterior distribution, where the density is more concentrated around the mean linked to the highest weight.

As an additional proof that the conditional Jeffreys' prior for the means of a Gaussian mixture model is improper, consider that, if the location or the scale parameters of a mixture model are unknown, this makes the model a location or a scale model, for which the Jeffreys' prior is improper in the location and the log-scale parameters, respectively (see [9] for details). In this case also the conditional Jeffreys' prior for the standard deviations when all the other parameters are considered known is improper.

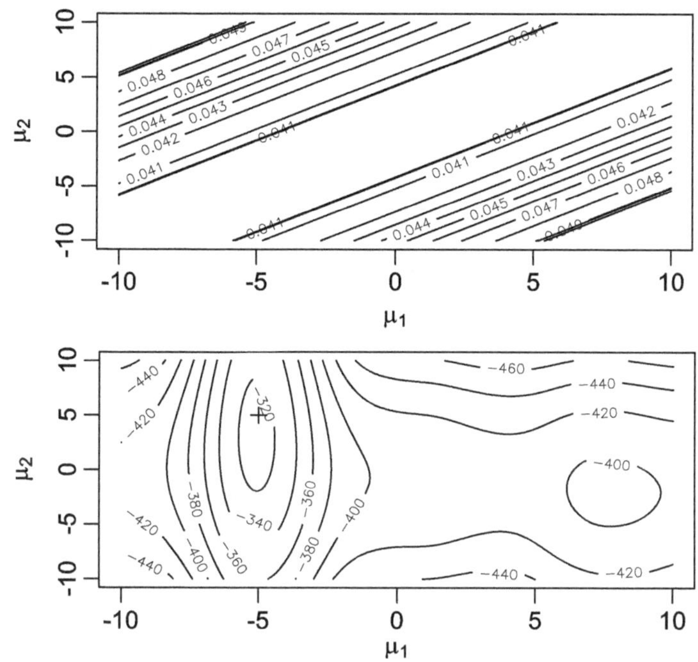

Fig. 4.3 Approximations of the (conditional) prior (*top*) and derived posterior (*bottom*) distributions for the means of the two-component Gaussian mixture model $0.5 \cdot N(-5,1) + 0.5 \cdot N(5,10)$

4.3 Implementation

Each element of the Fisher information matrix is an integral of the form presented in Eq. (4.2) which has to be approximated. We have applied both numerical integration and Monte Carlo integration and the results show that, in general, numerical integration obtained via Gauss–Kronrod quadrature yields more stable results. Nevertheless, when one or more proposed values for the standard deviations or the weights are too small, the approximations tend to be very dependent on the bounds used for numerical integration (usually chosen to omit a negligible part of the density). In this case Monte Carlo integration seems to yield more stable approximations and thus is applied by us. However, in these situations the approximation could lead to a negative determinant of the Fisher information matrix, even if it was very small in absolute value (of order 10^{-25} or even smaller). In this case, we have chosen to recompute the approximation until we get a positive number.

The computing expense due to deriving the Jeffreys' prior for a set of parameter values is $O(d^2)$, where d is the total number of (independent) parameters. A way to accelerate the Metropolis–Hastings algorithm used to approximate the posterior distribution derived from the Jeffreys' prior is the Delayed Acceptance algorithm proposed by Banterle et al. [1] (Algorithm 1).

Algorithm 1 Delayed Acceptance algorithm

Choose the initial values w^0, μ^0, σ^0
for i in $1:N$ **do**
  Propose $w^{prop}, \mu^{prop}, \sigma^{prop} \sim K(\cdot,\cdot,\cdot|w^{(i-1)}, \mu^{(i-1)}, \sigma^{(i-1)})$
  Simulate $u_1 \sim Unif(0,1)$ and $u_2 \sim Unif(0,1)$
  if $u_1 < \frac{l(w^{prop},\mu^{prop},\sigma^{prop})}{l(w^{(i-1)},\mu^{(i-1)},\sigma^{(i-1)})} \frac{K(w^{(i-1)},\mu^{(i-1)},\sigma^{(i-1)}|w^{prop},\mu^{prop},\sigma^{prop})}{K(w^{prop},\mu^{prop},\sigma^{prop}|w^{(i-1)},\mu^{(i-1)},\sigma^{(i-1)})}$ **then**
    if $u_2 < \frac{\pi^J(w^{prop},\mu^{prop},\sigma^{prop})}{\pi^J(w^{(i-1)},\mu^{(i-1)},\sigma^{(i-1)})}$ **then** Set $(w^{(i)}, \mu^{(i)}, \sigma^{(i)}) = (w^{prop}, \mu^{prop}, \sigma^{prop})$
    else $(w^{(i)}, \mu^{(i)}, \sigma^{(i)}) = (w^{(i-1)}, \mu^{(i-1)}, \sigma^{(i-1)})$
    end if
  else Set $(w^{(i)}, \mu^{(i)}, \sigma^{(i)}) = (w^{(i-1)}, \mu^{(i-1)}, \sigma^{(i-1)})$
  end if
end for

This exact algorithm allows to compute the Jeffreys' prior (the more expensive part of the posterior distribution) only when a first accept/reject step depending on the likelihood ratio (less costly) leads to acceptance and reduces the computational time by about 80 % (from an average of about 113 h with standard Metropolis–Hastings algorithm to an average of about 32 h with the Delayed Acceptance version) for 10^6 simulations for a three-component Gaussian mixture model with all parameters unknown, with a decrease of the acceptance rates from about 35 % with the standard Metropolis algorithm to about 20 % with the Delayed Acceptance. In combination with a reduction in the acceptance rate, the Delayed Acceptance version of the Metropolis–Hastings algorithm also induces a reduction of the effective sample size of about 35 %. We have used an algorithm that is adaptive during the burn-in period such that it leads to an acceptance rate above 20 % and below 40 %.

The Delayed Acceptance algorithm is an ideal solution for this setting: the likelihood is cheap to evaluate while the prior distribution is not only demanding but also non-informative. Therefore, it should have a limited influence with respect to the data when computing the posterior distribution and thus an early rejection due only to the likelihood ratio should not worsen the MCMC performances. Nevertheless, attention must be paid when applying the algorithm; since the prior distribution is improper, when the likelihood function is concentrated near regions of the parameter space where the prior distribution diverges (for examples in the case of Gaussian mixture models if the likelihood function is concentrated not far from values of standard deviations near 0), even if the first step based on the likelihood alone accepts a move, the first derivative of the prior distribution in that point may be too high in absolute value to allow the acceptance of the move and, therefore, the chain may be stuck or accept the proposed value only if the move is towards regions of even higher prior density.

A solution to this problem may be seen in splitting the likelihood ratio and using a part of it (relative to a small set of observations) jointly with the prior ratio in the second step. How many observations one has to consider in this splitting depends on the problem at hand and on the total number of observations.

4.4 The Posterior Distribution for a Mixture Model when Jeffreys' Prior is Used

It is a well-established fact in the literature that using independent improper priors for mixture models leads to improper posterior distributions, in particular if one of the components of the mixture model is not represented in the observed sample (i.e., there are no observations relative to at least one of the components of the mixture). One may use improper priors in mixture models by introducing some form of dependence between the components, as shown in [7]. Actually, the Jeffreys' prior does that by considering the Fisher information matrix. Checking for properness of the posterior distribution is unfeasible in an analytic way and the outcome of the classical Metropolis–Hastings algorithm and the version described in Sect. 4.3 targeting the posterior distribution derived from using the Jeffreys' prior has to be exploited in order to collect evidence that the posterior distribution is proper, even if the results which will be presented are not a conclusive proof of that.

4.4.1 Output of the MCMC Algorithm

Through extensive simulation, we have seen that for a sufficiently big sample size (at least equal to 20 for a three-component Gaussian mixture model) the MCMC chain never diverged. In particular, when the sample size decreases by up to ten data points, the component with the lowest weight is usually not identified and the chains may even get stuck, in particular because values of standard deviations close to 0 are accepted. The uncertainty on the posterior estimates depends on how variable that particular component is and how big the corresponding weight is. The acceptance rate of the MCMC algorithm is around 20 % for high sample sizes and it increases as the sample size decreases, unless the chain gets stuck (this happens with very low sample sizes).

4.4.2 An Example

Experiments with simulated data have been performed with different numbers of Gaussian components with different location and scale parameters generating the data and with different weights. The results are always similar, except for the fact that the uncertainty on the Bayesian estimates of the parameters increases as the components are closer, in terms of location. In particular, the Bayesian estimates of the components with the highest variances and/or the lowest weights are more variable.

Figures 4.4, 4.5, and 4.6 show the trace plots and the histograms of the MCMC chains approximating the marginal posterior distributions of the parameters of the three-component Gaussian mixture model

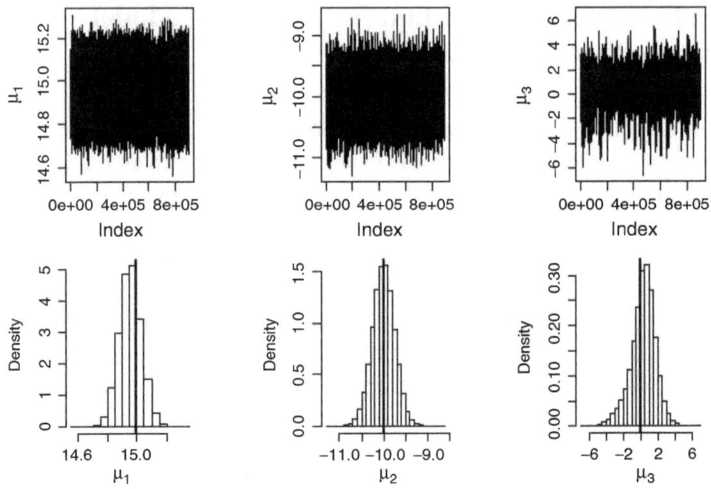

Fig. 4.4 Marginal posterior distributions (chains obtained via Metropolis–Hastings algorithm) for the means of a three-component Gaussian mixture model $0.65 \cdot N(15, 0.5) + 0.25 \cdot N(-10, 1) + 0.10 \cdot N(0, 5)$

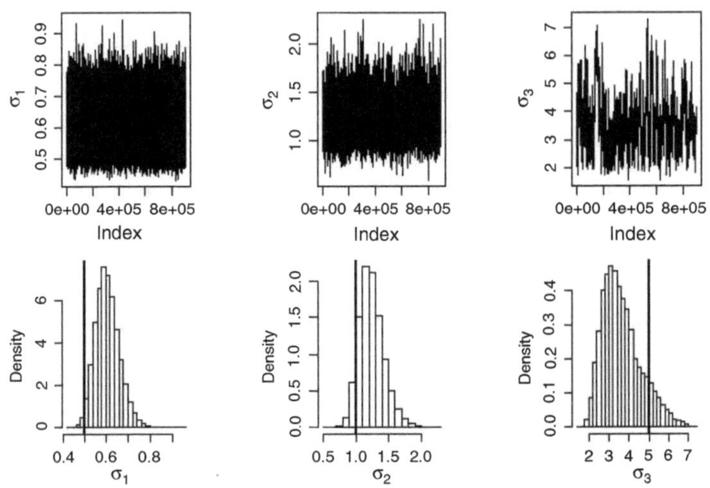

Fig. 4.5 Marginal posterior distributions for the standard deviations of the same model as in Fig. 4.4

$$0.65 \cdot N(15, 0.5) + 0.25 \cdot N(-10, 1) + 0.10 \cdot N(0, 5) \tag{4.13}$$

and for a sample size equal to 100. They show that the chains have reached convergence, with a higher uncertainty when estimating the mean and the standard deviation of the third component, being the one with the highest variability and the smallest weight.

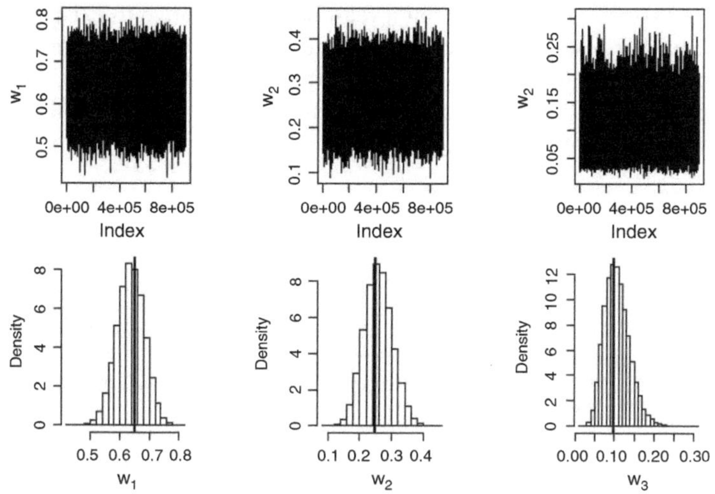

Fig. 4.6 Marginal posterior distributions for the weights of the same model as in Fig. 4.4

4.5 Discussion

The aim of this work is to propose an objective Bayesian analysis for the mixture model setting. The literature on mixture models shows that attention must be paid when using improper priors with these models and a generally accepted default solution does not exist. Nonetheless, we try to introduce a new objective Bayesian approach to handling mixture models by applying a Jeffreys' prior which considers the parameters of the dependent model. The prior has been shown to be improper, nevertheless extensive simulation studies suggest that the posterior could be proper, at least for a sufficiently high number of observations.

There are two important drawbacks when using the Jeffreys' prior for mixture models. First, the prior depends on integrals which have to be approximated. Possible solutions to this problem have been investigated, but they are beyond the scope of this paper. One may refer to [1] for an algorithm which may reduce the computational time of approximating the posterior distribution. Another solution is to reparameterize the model, as proposed by Mengersen and Robert [7], and exploit its independence features to reduce the dimension of the matrix to approximate. Second, the posterior distribution cannot be managed in an analytic way. Nevertheless there is no assurance that a distribution is proper if the Markov chain simulated via MCMC and used to approximate the posterior seems to converge. Future work will be aimed at the study of the relationship between the prior distribution and the likelihood function (in particular, the tails of the two functions) that makes the posterior proper in the setting of mixture models.

For the moment a widely accepted objective Bayesian approach in the setting of mixture models does not exist. This work could be seen as a way to understand if the Jeffreys' prior could represent a reasonable alternative to existing solutions. However, further research is needed, in particular to prove if and in which cases the posterior distribution derived from the Jeffreys' prior is proper and to generalize the Jeffreys' prior to models with non-Gaussian components or with a non-fixed number of components.

References

[1] Banterle, M., Grazian, C., Robert, C.P.: Accelerating Metropolis-Hastings algorithms: delayed acceptance with prefetching. arXiv:1406.2660 (2014)
[2] Basford, K.E., McLachlan, G.J.: Likelihood estimation with normal mixture models. Appl. Stat. **34**(3), 282–289 (1985)
[3] Diebolt, J. and Robert, C.P.: Estimation of finite mixture distributions through Bayesian sampling. J. R. Stat. Soc. Ser. B Stat Methodol. **56**(2), 363–375 (1994)
[4] Frühwirth-Schnatter, S.: Finite Mixture and Markov Switching Models: Modeling and Applications to Random Processes, 1st edn. Springer, New York (2006)
[5] Jeffreys', H.: Theory of Probability, 1st edn. The Clarendon Press, Oxford (1939)
[6] McLachlan, G.J., Peel, D.: Finite Mixture Models, 1st edn. Wiley, Newark (2000)
[7] Mengersen, K., Robert, C.: Testing for mixtures: a Bayesian entropic approach (with discussion). In: Berger, J., Bernardo, J., Dawid, A., Lindley, D., Smith, A. (eds.) Bayesian Statistics, vol. 5, pp. 255–276. Oxford University Press, Oxford (1996)
[8] Richardson, S., Green, P.J.: On Bayesian analysis of mixtures with an unknown number of components. J. R. Stat. Soc. Ser. B Stat Methodol. **59**(4), 731–792 (1997)
[9] Robert, C.: The Bayesian Choice, 2nd ed. Springer, New York (2001)
[10] Roeder, K., Wasserman, L.: Practical Bayesian density estimation using mixtures of normals. J. Am. Stat. Assoc. **92**(439), 894–902 (1997)
[11] Titterington, D.M., Smith, A.F.M., Makov, U.E.: Statistical Analysis of Finite Mixture Distributions, vol. 7. Wiley, Newark (1985)
[12] Wasserman, L.: Asymptotic inference for mixture models using data-dependent priors. J. R. Stat. Soc. Ser. B Stat Methodol. **62**(1), 159–180 (2000)

Chapter 5
A Subordinated Stochastic Process Model

Ana Paula Palacios, J. Miguel Marín, and Michael P. Wiper

Abstract We introduce a new stochastic model for non-decreasing processes which can be used to include stochastic variability into any deterministic growth function via subordination. This model is useful in many applications such as growth curves (children's height, fish length, diameter of trees, etc.) and degradation processes (crack size, wheel degradation, laser light, etc.). One advantage of our approach is the ability to easily deal with data that are irregularly spaced in time or different curves that are observed at different moments of time. With the use of simulations and applications, we examine two approaches to Bayesian inference for our model: the first based on a Gibbs sampler and the second based on approximate Bayesian computation (ABC).

Key words: ABC, Gibbs sampling, Growth models, Stochastic processes, Subordination

5.1 Introduction

Growth processes are usually described using discrete time models where the mean function is deterministic and a stochastic element is introduced via an additive, random noise component. An alternative approach is to consider continuous time modelling. In the literature some stochastic growth models are proposed using stochastic differential equations to model the variations ([1, 3]). However, the solution of these equations is not monotonically increasing and therefore can fail to model non-decreasing growth process like, for example, children's height, fish size or crack length. In this work, we introduce a new stochastic model for non-

A.P. Palacios (✉)
School of Computing and Mathematics, Plymouth University, Plymouth, UK
e-mail: ana.palacios@plymouth.ac.uk

J.M. Marín • M.P. Wiper
Department of Statistics, Universidad Carlos III de Madrid, Madrid, Spain
e-mail: jmmarin@est-econ.uc3m.es; m.wiper@est-econ.uc3m.es

S. Frühwirth-Schnatter et al. (eds.), *Bayesian Statistics from Methods to Models and Applications*, Springer Proceedings in Mathematics & Statistics 126,
DOI 10.1007/978-3-319-16238-6_5

decreasing processes that overcomes the problem described above. This model can be used to include stochastic variability into any deterministic growth function via subordination. That is, starting from a base process we construct a new process by applying a time change. Let X_t be a base stochastic process and T_t be the time change process (subordinator). The time-changed (subordinated process) Y_t is defined as $Y_t = X_{T_t}$. The main features of this model are that its paths are non-decreasing everywhere and, as a particular case, the mean function of the process is equal to the deterministic growth function used as time change.

5.2 The Model

The growth model that we propose is built upon a homogeneous, continuous time Markov process. It is commonly observed that growth in biological processes, and the wear in degradation processes, does not occur continuously. In contrast, growth or damage occurs per intervals of time. Furthermore, the growth velocity can also fluctuate. To represent this discontinuous growth, we start our model with a time homogeneous Markov process $\{U_t : t \geq 0\}$ with finite state space $\mathscr{S}$. States are ordered and they represent different levels of the growth rate. Transitions can only occur between neighbours. That is, if at time t the process is in state i, after an exponential amount of time it moves to one of the neighbouring states $i \to i+1$ or $i \to i-1$. This allows us to represent possible fluctuations in the growth rate but without abrupt changes. The process U_t is uniquely determined by the generator matrix, $\mathbf{Q}$, and the initial distribution of the process, ν_0. The transition rate matrix $\mathbf{Q}$ is a tridiagonal matrix

$$\mathbf{Q} = \begin{pmatrix} -\alpha & \alpha & 0 & 0 \; 0 & \dots \\ \beta & -(\alpha+\beta) & \alpha & 0 \; 0 & \dots \\ 0 & \beta & -(\alpha+\beta) & \alpha \; 0 & \dots \\ \vdots & \vdots & \vdots & \vdots \; \vdots & \\ 0 & 0 & 0 & 0 \; \beta & -\beta \end{pmatrix}$$

with parameters $\alpha > 0$, the instantaneous upward jump rate and $\beta > 0$, the instantaneous downward jump rate.

Now we define a continuous state process, $\{V_t : t \geq 0\}$, such that

$$V_t = \int_0^t U_s ds. \tag{5.1}$$

This is a non-decreasing, continuous time process which, being the integral of the growth rate, represents the total growth. Realisations of V_t are the path integrals of a simple stochastic process and their trajectories are piece-wise linear.

Beyond the growth fluctuations described earlier, most of these processes are also characterised by growth stages. For example, many growth processes show an

S-shape curve, where an initial (almost) steady period is followed by an exponential growth before a deceleration of the growth rate. We introduce these different stages in the model by manipulating the time (see for time change and subordination [5]), e.g. when the growth is slow, we want time to slow down but when the growth is exponential we want time to speed up. Thus, the time velocity will be governed by a deterministic non-decreasing function. We define our stochastic growth process, $\{Y_t : t \geq 0\}$ to be a continuous time stochastic process with continuous state space, defined as

$$Y_t = V_{T(t;\boldsymbol{\theta})}, \tag{5.2}$$

where $T(t;\boldsymbol{\theta})$ is any deterministic non-decreasing function of time and $\boldsymbol{\theta}$ is the vector of parameters.

It can be shown that, assuming a stationary state of the Markov process, the mean function of the process Y_t is proportional to the time change function T: $E[Y_t] = \mu T(t;\boldsymbol{\theta})$, where μ is the constant mean of the Markov process in the stationary state. As a particular case, if a proportion J of the function T is used as a time change and $J = 1/\mu$, then $E[Y_t] = T(t;\boldsymbol{\theta})$. This fact suggests to use as time change function any standard parametric function commonly used to model the phenomena of interest. For example, for population growth a logistic function could be used; for fish size the von Bertalanffy growth function could be used, etc. Figure 5.1 shows, on the right, 20 replications of experimental bacterial growth curves (the data used in this paper is a small subset of [2, 6]), and on the left, 20 realisations of the process Y_t when using the Gompertz function as time change. The real data consists of optical density measurements of bacterial cultures where the absorbance measurements were conducted at a wavelength of 595 nm.

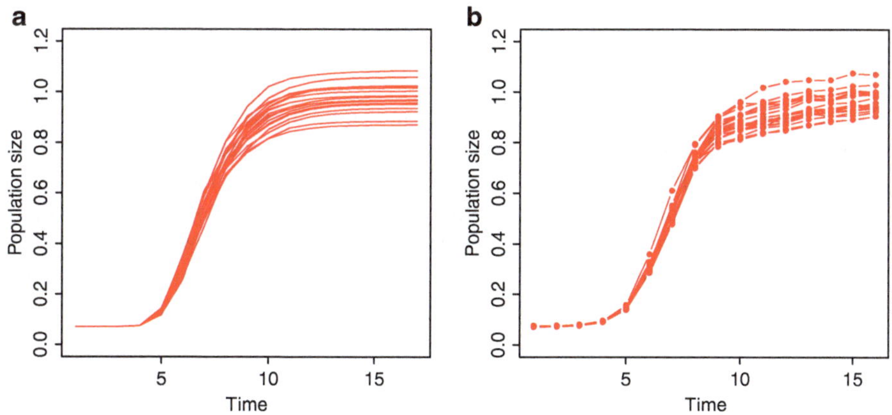

Fig. 5.1 Simulated realisations and real bacterial growth curves. (**a**) Several realisations, (**b**) Replications

The variance of the process V_t is linearly increasing with time and its magnitude depends on the instantaneous intensity rates of the Markov process. The greater the intensity rates, the lower the variance.

5.3 Bayesian Inference

The parameters of the model are the intensity rates α and β of the Markov process and the parameters $\boldsymbol{\theta}$ of the time change function. For this full model, the likelihood function $f(y|\alpha,\beta,\boldsymbol{\theta})$ is analytically unavailable which implies that frequentist approaches are infeasible. This limitation restricts the computational inference options to likelihood-free methods and we will illustrate this approach with a naive ABC example later in this section. However, for the simpler model with a Markov process U_t with two states $\{0,1\}$ and equal upward and downward jump rates, we obtain an explicit expression for the likelihood when conditioning on the initial state and the number of jumps in successive time intervals. Assume that we observe the growth data $y_{t_1} < \ldots < y_{t_n}$ at a sequence of time points, $0 < t_1 < \ldots < t_n$, say. The difference $\bar{y}_i = y_{t_i} - y_{t_{i-1}}$ is equal to the total time spent in state 1 during the transformed time interval $T_i = T(t_i) - T(t_{i-1})$. Conditioning on the number of jumps N_i in interval i, it is possible to show that the distribution of $\bar{y}_i$ is equal to the distribution of the order statistics of a uniform $(0,T_i)$ distribution where T_i is the length of interval i. Then, the conditional likelihood of $\bar{y}_i$ follows a scaled beta distribution and this allows for the implementation of a Gibbs sampling algorithm (see [4]). In what follows, we will discuss in more detail the implementation of two Bayesian approaches, namely Gibbs sampling and the ABC algorithm.

5.3.1 Gibbs Sampling

For the simple model with two states $\{0,1\}$ and equal upward and downward jump rates $\lambda = \alpha = \beta$, the parameters to be estimated are the intensity rate λ and the parameters of the time change function $\boldsymbol{\theta}$. In addition, the model has two latent variables, namely the initial state of the Markov process s_0 and the number of jumps per intervals N_i, for $i = 1,\ldots,n$. We implement the inference approach in two stages: in a first stage, the parameters of the time change function are estimated and, assuming them known in the second stage, the intensity rate is estimated.

Suppose that for each interval i we know the state of the underlying Markov chain at the starting time T_{i-1}, that is s_{i-1} and the number of jumps, N_i. Note that, given the number of jumps per interval and the initial state s_0, the successive s_i can be calculated as follows:

$$s_i = \text{mod}(s_{i-1} + N_i, 2), \quad i = 1, \ldots, n,$$

where mod(a,b) represents a modulo b. Suppose that we know the initial state, say s_0, at the start of the first time interval, and the number of state changes that occur in each time interval, say N_i, for $i = 1, \ldots, n$. Then, the likelihood function is:

$$f(\bar{\mathbf{y}}|\lambda, s_0, \mathbf{N}) = \prod_{i=1}^{n} f(\bar{y}_i|\lambda, s_{i-1}, N_i), \tag{5.3}$$

where $\mathbf{N} = (N_1, \ldots, N_n)$. The densities of each $\bar{y}_i$ are conditionally independent given the state at the start of interval i and the number of state transitions in the interval. Moreover, given the number of jumps, the likelihood function is conditionally independent of λ. Now consider two cases: when N_i is odd and when N_i is even. The analytical form of the conditional likelihood for these two cases is as follows:

- If N_i is odd, then the process spends half of its time intervals in state 1 and the remainder in state 0. Therefore, the distribution of the sum of $(N_i + 1)/2$ intervals is equal to the distribution of the order statistic $U_{((N_i+1)/2)}$, i.e. a scaled beta distribution

$$f(\bar{y}_i|N_i) = \frac{1}{B(\frac{N_i+1}{2}, \frac{N_i+1}{2})} \frac{\bar{y}_i^{(N_i+1)/2-1}(T_i - \bar{y}_i)^{(N_i+1)/2-1}}{T_i^{N_i}},$$

  where B denotes the beta function.
- If N_i is even, then the number of time intervals in period i is odd. Therefore, the process spends $N_i/2 + 1$ time intervals in state 1 if the initial state is 1, or $N_i/2$ if the initial state is 0. That is,

$$f(\bar{y}_i|s_{i-1}, N_i) = \frac{1}{B(\frac{N_i}{2} + s_{i-1}, \frac{N_i}{2} + (1 - s_{i-1}))} \frac{\bar{y}_i^{N_i/2+s_{i-1}-1}(T_i - \bar{y}_i)^{N_i/2-s_{i-1}}}{T_i^{N_i}}.$$

The conditional posterior distributions of λ and s_0 can be derived analytically. However, the posterior distribution of N_i does not have a closed form, and a Metropolis–Hasting step is necessary. Assuming that λ has a gamma prior distribution, say $\lambda \sim Gamma(a,b)$, then we have

$$\begin{aligned}
f(\lambda|\bar{\mathbf{y}}, s_0, \mathbf{N}) &\propto f(\bar{\mathbf{y}}|\lambda, s_0, \mathbf{N}) f(\lambda|s_0, \mathbf{N}) \\
&\propto f(\mathbf{N}|\lambda) f(\lambda), \\
\lambda|\bar{\mathbf{y}}, s_0, \mathbf{N} &\sim Gamma(a + n\bar{N}, b + n\bar{T}),
\end{aligned}$$

where $\bar{N} = (1/n)\Sigma_{i=1}^{n} N_i$ and $\bar{T} = (1/n)\sum_{i=1}^{n}(T_i - T_{i-1})$ is the average length of the transformed time intervals. Assuming that s_0 has a Bernoulli prior distribution with $P(s_0 = 1) = p$, then

$$P(s_0 = 1|\bar{\mathbf{y}}, \lambda, \mathbf{N}) = \frac{f(\bar{\mathbf{y}}|\lambda, s_0 = 1, \mathbf{N})P(s_0 = 1)}{f(\bar{\mathbf{y}}|\lambda, s_0 = 1, \mathbf{N})P(s_0 = 1) + f(\bar{\mathbf{y}}|\lambda, s_0 = 0, \mathbf{N})P(s_0 = 0)}.$$

This is straightforward to sample using a Gibbs step by computing analytically the conditional posterior probabilities of $s_0 = 0$ and $s_0 = 1$.

The posterior distributions of the N_is do not have a simple closed form and we use a Metropolis–Hastings algorithm to sample from these distributions based on generating candidate values from a Poisson distribution centred at the current value plus 0.5.

To illustrate this approach, a data set was simulated for given values of the time change function. For this example, we have chosen the Gompertz function as time change because this is one of the most common and successful parametric models used to describe bacterial growth. There are several parameterisations of the Gompertz function. In this work we use:

$$G(t) = D\exp\left(-\exp\left(1 + \frac{\mu e(\alpha - t)}{D}\right)\right),$$

where D is the distance between the maximum and the initial population size, μ is the maximum growth rate and α is a parameter that describes the lag period of the bacteria to start the binary fission. Assuming $\lambda = 10$, five replications were generated with 20 observations per curve. Gibbs sampling was performed to estimate the intensity rate and the results are shown in Fig. 5.2. The Markov chain has converged and the posterior distribution of λ is centred around the true value. The posterior mean is equal to 9.42 and the median is 9.38. The 95 % credible interval is equal to $(7.48, 11.51)$.

5.3.2 *Naive ABC Example*

For the more general case, with multiple states in the Markov chain, the previous Gibbs sampling cannot be implemented since it is restricted to the case of only two states in the Markov chain. Instead, likelihood-free methods must be used for the general case of a state space $\mathscr{S} = \{s_1, \ldots, s_k\}$. We illustrate this approach with a naive ABC implementation.

The ABC algorithm proceeds by sampling parameter values from the prior distributions and then simulating growth curve data (at the same time points as the observed data) given these parameter values. The general algorithm is as follows:

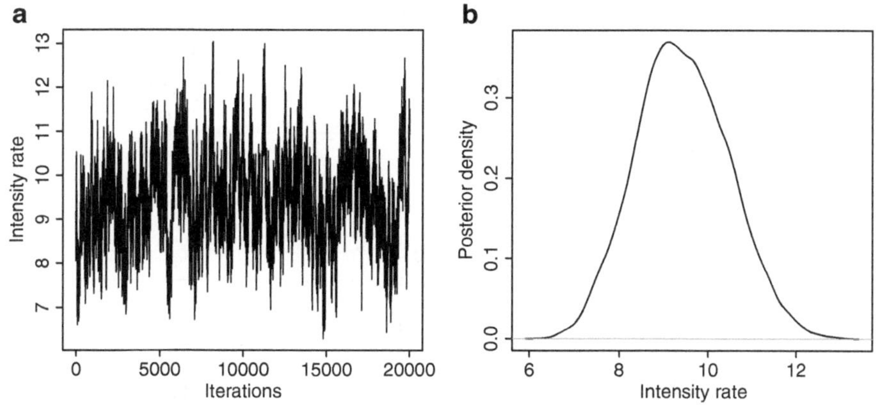

Fig. 5.2 Plots illustrating the Gibbs sampling outcomes. (**a**) Trace plot of the intensity rate, (**b**) Posterior distribution of the intensity rate

1. Sample from the prior distributions the intensity rate $\lambda^{(i)}$ and the time change parameters $\boldsymbol{\theta}^{(i)}$
2. Simulate data given $\lambda^{(i)},\boldsymbol{\theta}^{(i)}$: $\mathbf{y}^{(i)} \sim f(\mathbf{y}|\lambda^{(i)},\boldsymbol{\theta}^{(i)})$
3. Reject $\lambda^{(i)},\boldsymbol{\theta}^{(i)}$ if $d(\mathbf{y}^{(i)},\mathbf{y}) \geq \varepsilon$, where $d(\mathbf{y}^{(i)},\mathbf{y}) = \sum_{j=1}^{n} w_j |y_j^{(i)} - y_j|$
4. Repeat 1-3 until required number of candidates is accepted.

For the case where a single curve is observed, the weights are set to be equal. When the observed data consist of several replications of the same process, simulated data are evaluated by taking into account its distance with respect to the mean curve. The weights account for the variability of the process at different points of time. After this, we accept the 1 % of the generated parameter values with the lowest distance between the observed and simulated data.

This simplest rejection-ABC algorithm described above was implemented and applied to the real bacterial data of Fig. 5.1. The time change function chosen is again the Gompertz function. The parameters estimated are the intensity rate and the parameters of the Gompertz function. Informative prior distributions were assumed for the Gompertz parameters to avoid producing excessively many unreasonable parameter values. We compared one to one the simulated data with the mean observed data and summed over all the distances for each observation in a curve. When computing the distances between the simulated and observed data $|y_j^{(i)} - y_j|$ we penalised more for departures at earlier times (when less variability between curves is normally observed). To set the number of iterations required we first perform a simulation study. We found that increasing the number of iterations beyond 100,000 did not improve the accuracy (measured through the relative mean integrate square error of the parameters). Finally, we kept the best 1 % of the 100,000 sampled parameter values, i.e. the ones which minimise the distance. Several state spaces were considered and Table 5.1 summarises the results showing the estimated

Table 5.1 Estimated parameter values given different state spaces

	$\mathscr{S}=\{0.8,1\}$	$\mathscr{S}=\{0.8,1,1.2\}$	$\mathscr{S}=\{0.8,1,1.2,1.4\}$	$\mathscr{S}=\{0.8,0.9\}$	$\mathscr{S}=\{0.8,0.9,1\}$
λ	0.96	1.00	0.99	1.00	1.00
μ	0.28	0.28	0.27	0.28	0.26
α	3.88	3.90	3.87	3.89	3.80
D	1.07	1.03	1.04	1.09	1.10

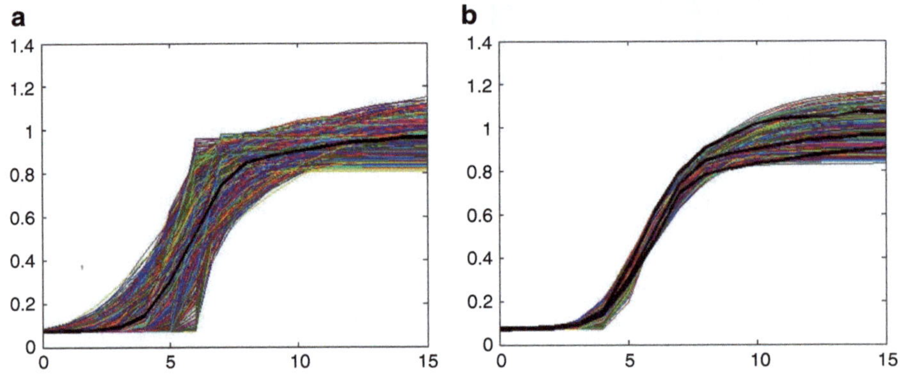

Fig. 5.3 Plots illustrating the ABC algorithm's outcomes. (**a**) Whole set of generated curves. The *thick black line* represents the mean observed curve (determined point-wise), (**b**) The best 1 % from the ABC sampler. *Thick black lines* represent the maximum, the minimum and the mean observed curves (determined point-wise)

posterior mean values for the parameters of the model. The posterior mean values for the Gompertz parameter were all biologically reasonable. In addition, Fig. 5.3 illustrates the curves simulated for the case with state space $\mathscr{S}=\{0.8,1\}$.

5.4 Conclusions

The aim of our work was to propose a new stochastic model suitable for growth processes and degradation data. Thus, the model developed has two desirable nice features. First, the growth paths are non-decreasing making the model feasible for a wide variety of applications such as crack size, fish or human growth. Second, as a particular case of the model, the mean function of the process is equal to the parametric function governing the time change. Another advantage of our approach is its ability to easily deal with data that are irregularly spaced in time or different curves that are observed at different moments of time. Finally, we have shown with the use of simulations and applications, two possible Bayesian approaches to fit the model, namely Gibbs sampling and approximate Bayesian computation. Gibbs sampling produces very good estimates for the intensity rate, assuming that the parameters of the time change function are known. Interesting extensions could be done with a full Gibbs sampling approach where all parameters are estimated simultaneously. In this way, the problem of underestimation of the total uncertainty present in two-stage inference can be overcome. In addition, the main limitation of this approach is that it is applicable only to models that assume only two states for the Markov chain. Alternatively, likelihood-free methods could be applied and a naive example using ABC was shown. This toy example required informative prior distributions for efficient estimation. More efficient alternatives like MCMC-ABC or sequential-ABC could be applied to allow for non-informative prior distributions.

References

[1] Donnet, S., Foulley, J., Samson, A.: Bayesian analysis of growth curves using mixed models defined by stochastic differential equations. Biometrics **66**, 733–741 (2010)
[2] Palacios, A.P., Marín, J.M., Quinto, E.J., Wiper, M.P.: Supplement to Bayesian modeling of bacterial growth for multiple populations. Ann. Appl. Stat. **8**, 1516–1537 (2014). doi:10.1214/14-AOAS720SUPPA
[3] Rensaw, E.: Stochastic Population Processes: Analysis, Approximations, Simulations. Oxford University Press, Oxford (2011)
[4] Robert, C.P., Casella, G.: Monte Carlo Statistical Methods, 2nd edn. Springer, New York (2005)
[5] Sato, K.I.: Lévy Processes and Infinitely Divisible Distributions. Cambridge University Press, Cambridge (1999)
[6] Toni, T., Welch, D., Strelkowa, N., Ipsen, A., Stumpf, M.P.: Approximate Bayesian computation scheme for parameter inference and model selection in dynamical systems. J. R. Soc. Interface **6**(31), 187–202 (2009)

Chapter 6
Bayesian Variable Selection for Generalized Linear Models Using the Power-Conditional-Expected-Posterior Prior

Konstantinos Perrakis, Dimitris Fouskakis, and Ioannis Ntzoufras

Abstract The power-conditional-expected-posterior (PCEP) prior developed for variable selection in normal regression models combines ideas from the power-prior and expected-posterior prior, relying on the concept of random imaginary data, and provides a consistent variable selection method which leads to parsimonious selection. In this paper, the PCEP methodology is extended to generalized linear models (GLMs). We define the PCEP prior in the GLM setting, explain the connections to other default model-selection priors, and present various posterior representations which can be used for model-specific posterior inference or for variable selection. The method is implemented for a logistic regression example with Bernoulli data. Results indicate that the PCEP prior leads to parsimonious selection for logistic regression models, similarly to the case of normal regression. Current limitations in generalizing the applicability of PCEP and possible solutions are discussed.

Key words: Power-conditional-expected-posterior prior, Bayesian variable selection, Generalized linear models

6.1 Introduction

During the last years, research in Bayesian variable selection has focused on the choice of suitable and meaningful priors for the parameters of competing models. On the one hand, using proper diffuse priors is problematic since posterior model odds are highly sensitive to the prior variances owing to the

---

K. Perrakis (✉) • I. Ntzoufras
Department of Statistics, Athens University of Economics and Business, Athens, Greece
e-mail: kperrakis@aueb.gr; ntzoufras@aueb.gr

D. Fouskakis
Department of Mathematics, National Technical University of Athens, Athens, Greece
e-mail: fouskakis@math.ntua.gr

S. Frühwirth-Schnatter et al. (eds.), *Bayesian Statistics from Methods to Models and Applications*, Springer Proceedings in Mathematics & Statistics 126,
DOI 10.1007/978-3-319-16238-6_6

Jeffreys–Lindley–Bartlett paradox [1, 14]. On the other hand, the use of improper priors results in indeterminate posterior odds, since the unknown normalizing constants do not cancel out in the calculation of Bayes factors. Due to these issues, a large body of literature is devoted to developing default proper priors for "objective" Bayesian model selection methods. Important contributions include, among others, Zellner's g-prior [24] in its various forms [5], the fractional Bayes factor approach [17], the intrinsic Bayes factor and the intrinsic prior [2], the intrinsic variable selection method [3], and the expected-posterior prior [18]. More recently, interest lies on mixtures of g-priors, including the hyper-g prior [13] with its extensions to generalized linear models (GLMs) [20] and to economic applications [12].

A key issue in the procedure of forming sensible and compatible prior distributions is the concept of underlying "imaginary" data. This concept is directly related to the power-prior approach [10] initially developed for "historical" data. For instance, Zellner's g-prior can be expressed as a power-prior with a fixed set of imaginary data [10, 24] and, similarly, that is the case for any mixture of g-prior, with the difference that additional uncertainty is introduced to the volume of information that the imaginary data contribute. In addition, the mechanism of imaginary data forms the basis of the expected-posterior prior [18], which is a generalization of the intrinsic prior [2]. Recently, the power-prior and the expected-posterior prior approaches were combined and used for applications of variable selection in normal linear regression. Initially, Fouskakis et al. [8] introduced the power-expected-posterior (PEP) prior applying jointly the power-prior and the expected-posterior prior approaches to the regression parameters and the error variance. In Fouskakis and Ntzoufras [7] the power-conditional-expected-posterior (PCEP) prior is developed by combining the two approaches for the regression parameters conditional on the error variance.

In this paper, we further extend the use of the PCEP prior to applications in GLMs. We present how the PCEP is defined in the GLM framework, its interpretation and relationship with power-priors and g-priors, and we also examine various posterior representations for single-model inference as well as variable selection. Current results from a Bernoulli logistic regression example indicate that the PCEP approach results in parsimonious selection, similarly to normal regression variable selection problems [7]. We also discuss current limitations and possible solutions for extending the application of the PCEP prior to other commonly used distributions belonging to the exponential family.

6.2 Model Specification with PCEP in GLMs

In this section, we initially describe the concept of imaginary data and the connection between the power-prior and g-prior in GLMs. We proceed by defining the PCEP prior and explaining the advantages of using this approach. We conclude with three different representations of the posterior distribution under the PCEP prior which can be used for MCMC sampling.

6.2.1 Imaginary Data and Power-Prior

Let us consider a set of imaginary data $\mathbf{y}^* = (y_1^*, y_2^*, \ldots, y_{n^*}^*)^T$ of size n^*. Then, for any model M_ℓ with parameter vector θ_ℓ, likelihood $f_\ell(\mathbf{y}^*|\theta_\ell)$ and baseline prior $\pi_\ell^{\mathrm{N}}(\theta_\ell)$, we can obtain a "sensible" prior for the model parameters from

$$\pi_\ell^{\mathrm{N}}(\theta_\ell|\mathbf{y}^*, \delta) \propto f_\ell(\mathbf{y}^*|\theta_\ell)^{1/\delta} \pi_\ell^{\mathrm{N}}(\theta_\ell). \tag{6.1}$$

This is the power-prior of Ibrahim and Chen [10]. The parameter δ controls the weight that the imaginary data contribute to the "final" posterior distribution of θ_ℓ. For $\delta = 1$, (6.1) is exactly equal to the posterior distribution of θ_ℓ after observing the imaginary data $\mathbf{y}^*$. For $\delta = n^*$ the contribution of the imaginary data to the overall posterior is equal to one data point; i.e., a prior having a unit-information interpretation [11].

6.2.2 Relation of Power-Prior and *g*-Prior in GLMs

We focus on variable selection problems for generalized linear models (GLMs), i.e. for models M_ℓ with parameters $\theta_\ell = (\beta_\ell, \phi_\ell)^T$ and response data $\mathbf{y} = (y_1, \ldots, y_n)^T$ with likelihood given by

$$f_\ell(\mathbf{y}|\beta_\ell, \phi_\ell) = \exp\left(\sum_{i=1}^{n} \frac{y_i \vartheta_{\ell(i)} - b(\vartheta_{\ell(i)})}{a(\phi_{\ell(i)})} + \sum_{i=1}^{n} c(y_i, \phi_{\ell(i)}) \right),$$

$$\vartheta_{\ell(i)} = g \circ b'^{-1}\left(\mathbf{X}_{\ell(i)} \beta_\ell\right),$$

where $\mathbf{X}_\ell$ is an $n \times d_\ell$ design matrix and $g \circ b'^{-1}(\boldsymbol{\vartheta}_\ell)$ is the inverse function of $g \circ b'(\vartheta_\ell) = g\big(b'(\vartheta_\ell)\big)$, ϑ_ℓ and ϕ_ℓ are the location and dispersion parameters of the exponential family, respectively, $a(\cdot)$, $b(\cdot)$, $c(\cdot)$ are functions specifying the structure of the distribution, and $g(\cdot)$ is the link function connecting the mean of the response y_i with the linear predictor.

Under the power-prior approach in (6.1) the prior of β_ℓ conditional on the parameter ϕ_ℓ and the imaginary data $\mathbf{y}^*$ of size n^* is

$$\pi_\ell^{\mathrm{N}}(\beta_\ell|\phi_\ell, \mathbf{y}^*, \delta) \propto \exp\left(\sum_{i=1}^{n^*} \frac{y_i^* \vartheta_{\ell(i)} - b(\vartheta_{\ell(i)})}{\delta a(\phi_{\ell(i)})} \right) \pi_\ell^{\mathrm{N}}(\beta_\ell|\phi_\ell).$$

Assuming a reference baseline prior for β_ℓ, i.e. $\pi_\ell^{\mathrm{N}}(\beta_\ell|\phi_\ell) \propto 1$, then we have asymptotically

$$\widehat{\pi}_\ell^{\mathrm{N}}(\beta_\ell|\phi_\ell, \mathbf{y}^*, \delta) \approx f_{N_{d_\ell}}\left(\beta_\ell; \widehat{\beta}_\ell^*, \delta\big(\mathbf{X}_\ell^{*T} \mathbf{H}_\ell^* \mathbf{X}_\ell^*\big)^{-1}\right), \tag{6.2}$$

where $\widehat{\beta}_\ell^*$ is the MLE of β_ℓ for data $\mathbf{y}^*$ and design matrix $\mathbf{X}_\ell^*$, $\mathbf{H}_\ell^* = \text{diag}(h_{\ell(i)}^*)$, with $h_{\ell(i)}^{*-1} = \left(\frac{\partial g(\mu_{\ell(i)})}{\partial \mu_{\ell(i)}}\right)^2 a(\phi_{\ell(i)}) b''(\vartheta_{\ell(i)})$ and $\mu_{\ell(i)} = b'(\vartheta_{\ell(i)})$, and $f_{N_d}(\mathbf{x}; \mu, \Sigma)$ denotes the density of the d-dimensional normal distribution with mean μ and variance-covariance matrix Σ.

Recall that the extension of Zellner's g-prior for GLMs, according to the definition in [16, 20], is of the following form

$$\beta_\ell | \phi_\ell \sim N_{d_\ell}\Big(\beta_\ell ; \mu_\ell, g\big(\mathbf{X}_\ell^T \mathbf{H}_\ell \mathbf{X}_\ell\big)^{-1}\Big). \tag{6.3}$$

Thus, the Zellner's g-prior can be interpreted as a power-prior on imaginary data with δ having the role of g and $\mathbf{X}_\ell^* = \mathbf{X}_\ell$, $\mathbf{H}_\ell^* = \mathbf{H}_\ell$. The familiar zero-mean representation in (6.3), i.e. $\mu_\ell = \mathbf{0}$, arises when all imaginary data in (6.2) are the same, i.e. $\mathbf{y}^* = g^{-1}(0)\mathbf{1}_{n^*}$, and $a(\phi_{\ell(i)}) = \phi_\ell / w_i$, where $\mathbf{1}_{n^*}$ is a vector of ones of size n^* and w_i is a known fixed weight; for details, see [16, 20].

6.2.3 Power-Conditional-Expected-Posterior Prior

The PCEP prior is derived by combining the power-prior approach [10] and the expected-posterior prior approach [18]. Consider initially, the conditional-expected-posterior (CEP) prior given by

$$\pi_\ell^{\text{CEP}}(\beta_\ell, \phi_\ell) = \pi_\ell^{\text{CEP}}(\beta_\ell | \phi_\ell) \pi_\ell^{\text{N}}(\phi_\ell),$$

where $\pi_\ell^{\text{N}}(\phi_\ell)$ is the baseline prior for ϕ_ℓ and

$$\pi_\ell^{\text{CEP}}(\beta_\ell | \phi_\ell) = \int \pi_\ell^{\text{N}}(\beta_\ell | \phi_\ell, \mathbf{y}^*) m_0^{\text{N}}(\mathbf{y}^*) \mathrm{d}\mathbf{y}^*. \tag{6.4}$$

Prior (6.4) corresponds to the expected-posterior-prior [18] for β_ℓ conditional on ϕ_ℓ, with the imaginary data coming from the *marginal* or *prior predictive* distribution

$$m_0^{\text{N}}(\mathbf{y}^*) = \int f_0(\mathbf{y}^* | \beta_0, \phi_0) \pi_0^{\text{N}}(\beta_0 | \phi_0) \pi_0^{\text{N}}(\phi_0) \mathrm{d}\beta_0 \mathrm{d}\phi_0,$$

where f_0 and π_0^{N} are the likelihood and baseline prior, respectively, of the null model M_0. In this context the null model is used as a reference model; see discussion in Sect. 6.2.4. Applications of this prior to normal regression variable selection can be found in [6].

The PCEP prior, which was gradually developed in [7] and [8] for normal regression models, is constructed by raising the likelihood, involved in (6.4), to the power $1/\delta$, that is

$$\pi_\ell^{\text{PCEP}}(\beta_\ell, \phi_\ell | \delta) = \pi_\ell^{\text{PCEP}}(\beta_\ell | \phi_\ell, \delta) \pi_\ell^{\text{N}}(\phi_\ell), \tag{6.5}$$

where

$$\pi_\ell^{\text{PCEP}}(\beta_\ell|\phi_\ell,\delta) = \int \pi_\ell^{\text{N}}(\beta_\ell|\phi_\ell,\mathbf{y}^*,\delta) m_0^{\text{N}}(\mathbf{y}^*|\delta) \text{d}\mathbf{y}^* \tag{6.6}$$

and

$$\pi_\ell^{\text{N}}(\beta_\ell|\phi_\ell,\mathbf{y}^*,\delta) = \frac{f_\ell(\mathbf{y}^*|\beta_\ell,\phi_\ell,\delta)\pi_\ell^{\text{N}}(\beta_\ell|\phi_\ell)}{m_\ell^{\text{N}}(\mathbf{y}^*|\phi_\ell,\delta)}. \tag{6.7}$$

The likelihood involved in (6.7) is the *density-normalized* power likelihood, i.e.

$$f_\ell(\mathbf{y}^*|\beta_\ell,\phi_\ell,\delta) = \frac{f_\ell(\mathbf{y}^*|\beta_\ell,\phi_\ell)^{1/\delta}}{\int f_\ell(\mathbf{y}^*|\beta_\ell,\phi_\ell)^{1/\delta} \text{d}\mathbf{y}^*}, \tag{6.8}$$

while

$$m_\ell^{\text{N}}(\mathbf{y}^*|\phi_\ell,\delta) = \int f_\ell(\mathbf{y}^*|\beta_\ell,\phi_\ell,\delta)\pi_\ell^{\text{N}}(\beta_\ell|\phi_\ell) \text{d}\beta_\ell \tag{6.9}$$

is the prior predictive of $\mathbf{y}^*$ for model M_ℓ conditional on ϕ_ℓ and δ and

$$m_0^{\text{N}}(\mathbf{y}^*|\delta) = \int f_0(\mathbf{y}^*|\beta_0,\phi_0,\delta)\pi_0^{\text{N}}(\beta_0|\phi_0)\pi_0^{\text{N}}(\phi_0) \text{d}\beta_0 \text{d}\phi_0 \tag{6.10}$$

is the prior predictive of $\mathbf{y}^*$ for model M_0 conditional on δ.

6.2.4 PCEP Interpretation and Specification

As seen in (6.6), the PCEP prior is the average of the posterior of β_ℓ given $\mathbf{y}^*$ over the prior-predictive of a reference model. It is therefore an objective prior which introduces an extra hierarchical level to account for the uncertainty in the imaginary data. If we further take into account the normal approximation in (6.2), it can also be considered as a mixture of g-priors with a hyper-prior assigned to $\mathbf{y}^*$ rather than to the variance multiplicator g (which is here substituted by δ).

Nevertheless, the PCEP methodology still depends on the selection of the power parameter δ, the size n^* of the imaginary sample, and the choice of the reference model. Following [8], we recommend:

- to set $\delta = n^*$ so that the imaginary sample contributes information equal to one data point, leading to a unit-information interpretation [11].
- to set $n^* = n$ and consequently $\mathbf{X}_\ell^* = \mathbf{X}_\ell$; this way we dispense with the need of selecting and averaging over "minimal" training samples as in the intrinsic and expected-posterior prior approaches [2, 18]; see [8] for a detailed discussion.
- to use the constant model M_0 as the reference model in order to support a-priori the most parsimonious data-generating assumption; see [22] for a related discussion.

As illustrated in [7, 8] for normal regression models, power-expected-posterior priors result in a consistent variable selection methodology. In addition, variable selection through PCEP using a g-baseline prior results in more parsimonious models than the hyper-g and the g-prior [7].

6.2.5 Posterior Distribution Under PCEP in GLMs

In normal regression models the PCEP prior is a conjugate normal-inverse gamma distribution which leads to fast and simple computations, cf. [7]. For the rest of GLMs the integration involved for deriving the PCEP prior is intractable. However, we can use the hierarchical model, i.e. without marginalizing in (6.6), and sample from the joint posterior distribution of β_ℓ, ϕ_ℓ and $\mathbf{y}^*$. From (6.5), (6.6), and (6.7) we have that

$$\begin{aligned}\pi_\ell^{\text{PCEP}}(\beta_\ell,\phi_\ell,\mathbf{y}^*|\mathbf{y},\delta) &\propto f_\ell(\mathbf{y}|\beta_\ell,\phi_\ell)\pi_\ell^{\text{N}}(\beta_\ell|\phi_\ell,\mathbf{y}^*,\delta)\pi_\ell^{\text{N}}(\phi_\ell)m_0^{\text{N}}(\mathbf{y}^*|\delta)\\ &\propto f_\ell(\mathbf{y}|\beta_\ell,\phi_\ell)\frac{f_\ell(\mathbf{y}^*|\beta_\ell,\phi_\ell,\delta)\pi_\ell^{\text{N}}(\beta_\ell|\phi_\ell)}{m_\ell^{\text{N}}(\mathbf{y}^*|\phi_\ell,\delta)}\\ &\quad\times\pi_\ell^{\text{N}}(\phi_\ell)m_0^{\text{N}}(\mathbf{y}^*|\delta).\end{aligned}\tag{6.11}$$

A problem when working with non-normal GLMs is that the prior-predictive distributions $m_0^{\text{N}}(\mathbf{y}^*|\delta)$ and $m_\ell^{\text{N}}(\mathbf{y}^*|\phi_\ell,\delta)$, defined in (6.10), and (6.9) respectively, are not known in closed form. One solution is to use the Laplace approximation for both.

Alternatively, we can avoid estimating the prior predictive $m_0^{\text{N}}(\mathbf{y}^*|\delta)$ of the null model by augmenting the parameter space further and include the parameter vector $(\beta_0,\phi_0)^T$ of the null model in the joint posterior. Under this approach, we deduce that

$$\begin{aligned}\pi_\ell^{\text{PCEP}}(\beta_\ell,\phi_\ell,\beta_0,\phi_0,\mathbf{y}^*|\mathbf{y},\delta) &\propto f_\ell(\mathbf{y}|\beta_\ell,\phi_\ell)\frac{f_\ell(\mathbf{y}^*|\beta_\ell,\phi_\ell,\delta)\pi_\ell^{\text{N}}(\beta_\ell|\phi_\ell)}{m_\ell^{\text{N}}(\mathbf{y}^*|\phi_\ell,\delta)}\pi_\ell^{\text{N}}(\phi_\ell)\\ &\quad\times f_0(\mathbf{y}^*|\beta_0,\phi_0,\delta)\pi_0^{\text{N}}(\beta_0|\phi_0)\pi_0^{\text{N}}(\phi_0).\end{aligned}\tag{6.12}$$

If we further want to avoid the estimation of $m_\ell^{\text{N}}(\mathbf{y}^*|\phi_\ell,\delta)$, we can make use of the asymptotic result presented in (6.2) (note that we control the effect of the imaginary data) and work with the following expression

$$\begin{aligned}\pi_\ell^{\text{PCEP}}(\beta_\ell,\phi_\ell,\beta_0,\phi_0,\mathbf{y}^*|\mathbf{y},\delta) &\propto f_\ell(\mathbf{y}|\beta_\ell,\phi_\ell)\widehat{\pi}_\ell^{\text{N}}(\beta_\ell|\phi_\ell,\mathbf{y}^*,\delta)\pi_\ell^{\text{N}}(\phi_\ell)\\ &\quad\times f_0(\mathbf{y}^*|\beta_0,\phi_0,\delta)\pi_0^{\text{N}}(\beta_0|\phi_0)\pi_0^{\text{N}}(\phi_0),\end{aligned}\tag{6.13}$$

with $\widehat{\pi}_\ell^{\text{N}}(\beta_\ell|\phi_\ell,\mathbf{y}^*,\delta)$ as defined in (6.2).

Sampling from the posterior distributions presented in (6.11), (6.12), and (6.13) is possible with standard Metropolis-within-Gibbs algorithms, i.e. sampling sequentially each component from the full conditional distribution with Metropolis–Hastings steps. For commonly used GLMs, such as logistic or Poisson regression models, $\phi_\ell = 1$, which simplifies the algorithms. It is also worth noting that for $\phi_\ell = 1$ there is no distinction between the PEP prior developed in [8] and the PCEP prior [7]. Moreover, computations are simplified further when using a reference baseline prior, i.e. $\pi_\ell^{\mathrm{N}}(\beta_\ell|\phi_\ell) \propto 1$.

6.3 Variable Selection Under PCEP in GLMs

Here we present a variable selection technique under the PCEP prior based on the Gibbs variable selection (GVS) algorithm [4]. For simplicity of illustration we consider GLMs where $\phi_\ell = 1$. In addition, we have to introduce a slight change in notation for the model indicator, but note that there is a direct one-to-one correspondence with the previous notation.

GVS introduces a vector of binary indicators $\gamma \in \{0,1\}^p$ representing which of the p possible sets of covariates are included in a model. Assuming that the constant is always included, the linear predictor can be written as $\eta = \beta_0 + \sum_{j=1}^{p} \gamma_j \mathbf{X}_j \beta_j$. We partition the vector β into $(\beta_\gamma, \beta_{\setminus\gamma})$, corresponding to those components of β that are included ($\gamma_j = 1$) and excluded ($\gamma_j = 0$) from the model, and define the baseline prior of β and γ as

$$\pi^{\mathrm{N}}(\beta,\gamma) = \pi^{\mathrm{N}}(\beta|\gamma)\pi^{\mathrm{N}}(\gamma) = \pi^{\mathrm{N}}(\beta_\gamma|\gamma)\pi^{\mathrm{N}}(\beta_{\setminus\gamma}|\gamma)\pi^{\mathrm{N}}(\gamma),$$

where the actual baseline prior choice involves only β_γ, since $\pi^{\mathrm{N}}(\beta_{\setminus\gamma}|\gamma)$ is just a *pseudo-prior* used for balancing the dimensions between model spaces. For γ we use the hierarchical prior $\gamma|\tau \sim \mathit{Bernoulli}(\tau)$ and $\tau \sim \mathit{Beta}(1,1)$ in order to account for the appropriate multiplicity adjustment [21]. The resulting prior model probabilities are

$$\pi^{\mathrm{N}}(\gamma) = \frac{1}{p+1}\binom{p}{p_\gamma}^{-1},$$

where p is the total number of covariates and p_γ is the number of covariates that are included in model M_γ. Under the GVS setting the posterior distribution presented in (6.11) can be expressed as

$$\pi^{\mathrm{PCEP}}(\beta_\gamma,\beta_{\setminus\gamma},\gamma,\mathbf{y}^*|\mathbf{y},\delta) \propto f(\mathbf{y}|\beta_\gamma,\gamma)\frac{f(\mathbf{y}^*|\beta_\gamma,\gamma,\delta)\pi^{\mathrm{N}}(\beta_\gamma|\gamma)}{m_\gamma^{\mathrm{N}}(\mathbf{y}^*|\delta)} \times\pi^{\mathrm{N}}(\beta_{\setminus\gamma}|\gamma)\pi^{\mathrm{N}}(\gamma)m_0^{\mathrm{N}}(\mathbf{y}^*|\delta). \quad (6.14)$$

Some remarks concerning implementation of GVS for the posterior in (6.14) are:

- Commonly, a flat baseline prior is imposed on β_γ, therefore $\pi^{\mathrm{N}}(\beta_\gamma|\gamma)$ is eliminated from the corresponding expressions.
- A usual choice for the pseudo-prior of $\beta_{\setminus\gamma}$ is the product of independent normal distributions with means equal to the maximum-likelihood estimates of the full model and standard deviations equal to the standard errors of these estimates.
- $\beta_{\setminus\gamma}$ is sampled directly from the pseudo-prior and the γ_j's from the full conditional which is a simple Bernoulli distribution; see [4] for details.
- Sampling β_γ and $\mathbf{y}^*$ from the full conditionals requires Metropolis–Hastings steps.
- The full conditional of $\mathbf{y}^*$ depends on $m_0^{\mathrm{N}}(\mathbf{y}^*|\delta)$ and $m_\gamma^{\mathrm{N}}(\mathbf{y}^*|\delta)$ which are unknown; as discussed in Sect. 6.2.5, one can use the Laplace approximation or, alternatively, use the posterior distributions presented in (6.12) and (6.13).

A detailed description of the GVS algorithm, based on the posterior distribution in (6.12), for Bernoulli logistic regression can be found in the appendix.

6.4 Example: Bernoulli Logistic Regression

In this section we present results for a variable selection problem in a logistic regression application to Bernoulli data, namely the Pima Indians diabetes data set [19], which contains $n = 532$ complete records on diabetes presence associated with $p = 7$ covariates. The number of competing models is $2^7 = 128$. The particular dataset has been considered by many authors under various methods, e.g. hyper-g priors [20] and auxiliary variable selection methods [9, 23].

We adopt a flat baseline prior for the regression parameters, i.e. $\pi_\ell^{\mathrm{N}}(\beta_\ell) \propto 1$. Furthermore, based on the recommendations discussed in Sect. 6.2.4, we set $\delta = n^* = n = 532$. Initially, we should note that we implemented MCMC runs for the full model with the purpose of comparing the approaches described in (6.11), (6.12), and (6.13) based on the three different representations of the posterior distribution. Our results (not presented here) in terms of posterior summaries are almost identical.

Here, we focus on the results from variable selection using the GVS algorithm described in Sect. 6.3 combined with the augmented posterior in (6.12). Results are based on 41,000 iterations, using the first 1,000 for burn-in; the posterior marginal inclusion probabilities under PCEP are presented in Table 6.1. For comparison reasons, Table 6.1 also includes the corresponding estimates presented in [20] under BIC and three variations of the hyper-g prior, namely $\pi(g) = IG(1/2, n/2)$ [25], $n(g) = 1/[n(1+g/n)^2]$, i.e. a hyper- g/n prior with $a = 4$ [13], and $\pi(g) = IG(10^{-3}, 10^{-3})$, where $IG(a,b)$ is the inverse-gamma distribution with shape a and scale b. The estimates for the strongly influential covariates X_1, X_2, X_5, and X_6 are more or less the same under the four approaches. On the contrary, the posterior marginal inclusion probabilities for covariates X_3, X_4, and X_7 are much lower under PCEP in comparison with the ones under the hyper-g prior methods. The median probability model using PCEP and BIC is $X_1 + X_2 + X_5 + X_6$, while the three hyper-g

Table 6.1 Posterior marginal inclusion probabilities for the Pima Indians diabetes data

	Variables	PCEP	BIC[a]	Hyper-g priors[a]		
				$IG(1/2,n/2)$	$n(g)=1/[n(1+g/n)^2]$	$IG(10^{-3},10^{-3})$
X_1	Number of pregnancies	0.951	0.946	0.961	0.965	0.968
X_2	Glucose concentration	1.000	1.000	1.000	1.000	1.000
X_3	Diastolic blood pressure	0.136	0.100	0.252	0.309	0.353
X_4	Triceps skin fold thickness	0.139	0.103	0.248	0.303	0.346
X_5	Body mass index	0.997	0.997	0.998	0.998	0.998
X_6	Diabetes pedigree function	0.992	0.987	0.994	0.995	0.996
X_7	Age	0.390	0.334	0.528	0.586	0.629

[a] Results from [20]

methods also include variable X_7. These results suggest that the PCEP prior can lead to more parsimonious selection, which is in agreement with the results presented in [7, 8] for normal regression models. Boxplots of the posterior marginal inclusion probabilities and posterior model probabilities for the two best ranking models under PCEP, derived from splitting the posterior sample into 40 batches of size 1,000, are presented in Fig. 6.1. The maximum a-posteriori model using PCEP is the same as the median probability model in this example, with a posterior model probability around 0.50.

6.5 Generalizing the Applicability of the PCEP Prior

A problem which arises when using the PCEP prior in GLMs is that the normalized power-likelihood presented in (6.8) is not always of a known form or in some cases it is extremely inconvenient to work with. For the normal and Bernoulli distributions, the power-likelihood is of the same form, which considerably simplifies computations. This is also true for the exponential and the beta distributions, but not for all members of the exponential family.

For instance, for the popular binomial and Poisson regression models the normalized power-likelihood is a discrete distribution, and although it is feasible to evaluate it, the additional computational burden may render its implementation time consuming and inefficient. One possible solution would be the use of an

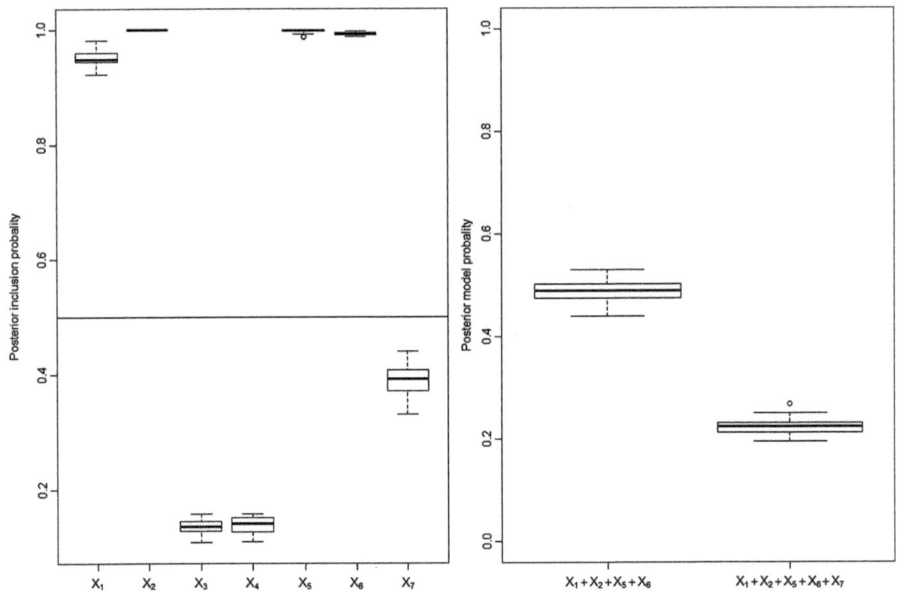

Fig. 6.1 Boxplots of the posterior marginal inclusion probabilities (*left*) and posterior model probabilities for the two best ranking models (*right*) under PCEP from 40 batches of size 1,000

exchange-rate algorithm for doubly intractable distributions [15]. A more general orientation is to redefine the PCEP prior, presented in Sect. 6.2.3, in the following ways: i) use the unnormalized power-likelihood for model M_ℓ but define the prior-predictive distribution of model M_0 based on the real likelihood (with δ=1) and ii) use the unnormalized power-likelihood both in M_ℓ and in M_0 and try to normalize the resulting PCEP prior.

6.6 Future Research

Our main aim is to extend the PCEP methodology to all members of the exponential family. To this end we are presently examining the PCEP methodology under the two variations discussed in Sect. 6.5 using unnormalized likelihoods. Current theoretical results for normal linear regression models show that the two variations of the PCEP prior have similar asymptotic behavior with the PCEP prior which is based on the normalized power-likelihood. It remains to be examined theoretically and/or empirically whether this result is also valid for non-normal GLMs.

Furthermore, for non-normal GLMs, we intend to examine two important characteristics, namely, the variance of the PCEP prior and the limiting behavior of the Bayes factor under PCEP. In normal regression the variance of the PCEP prior is larger than that of the g-prior leading to more parsimonious inference, while the asymptotic behavior of the Bayes factor is equivalent to that of the BIC criterion, thus resulting in a consistent variable selection method [7].

Further research directions under consideration include introducing an extra hierarchical level by assigning a hyper-prior on δ and also using shrinkage baseline priors (e.g. LASSO priors) with the purpose of extending the methodology to small n large p problems.

Acknowledgements This research has been co-financed in part by the European Union (European Social Fund-ESF) and by Greek national funds through the Operational Program "Education and Lifelong Learning" of the National Strategic Reference Framework (NSRF)-Research Funding Program: Aristeia II/PEP-BVS.

Appendix: GVS Implementation Under PCEP for Bernoulli Logistic Regression

Here we present the details concerning implementation of GVS based on the data-augmented posterior distribution in (6.12). Let d denote the total number of covariates, d_γ the number of active covariates ($\gamma_j = 1$), and $d_{\setminus\gamma}$ the number of inactive covariates ($\gamma_j = 0$). The posterior distribution is

$$\pi^{\mathrm{PCEP}}(\beta_\gamma, \beta_{\setminus\gamma}, \beta_0, \gamma, \mathbf{y}^*|\mathbf{y}, \delta) \propto f(\mathbf{y}|\beta_\gamma, \gamma)\frac{f(\mathbf{y}^*|\beta_\gamma, \gamma, \delta)\pi^{\mathrm{N}}(\beta_\gamma|\gamma)}{m_\gamma^{\mathrm{N}}(\mathbf{y}^*|\delta)}$$
$$\times \pi^{\mathrm{N}}(\beta_{\setminus\gamma}|\gamma)\pi^{\mathrm{N}}(\gamma)f(\mathbf{y}^*|\beta_0, \delta)\pi^{\mathrm{N}}(\beta_0).$$

Analytically, we have that:

- $f(\mathbf{y}|\beta_\gamma, \gamma) = \prod_{i=1}^{n}\left[p_{\gamma(i)}^{y_i}\left(1-p_{\gamma(i)}\right)^{1-y_i}\right]$, where $p_{\gamma(i)} = \frac{\exp\left(\mathbf{X}_{\gamma(i)}\beta_\gamma\right)}{1+\exp\left(\mathbf{X}_{\gamma(i)}\beta_\gamma\right)}$.
- $f(\mathbf{y}^*|\beta_\gamma, \gamma, \delta) = \prod_{i=1}^{n^*}\left[p_{\gamma(i)}^{*\,y_i^*}(1-p_{\gamma(i)}^*)^{1-y_i^*}\right]$, where $p_{\gamma(i)}^* = \frac{\left(p'_{\gamma(i)}\right)^{1/\delta}}{\left(p'_{\gamma(i)}\right)^{1/\delta}+\left(1-p'_{\gamma(i)}\right)^{1/\delta}}$ and $p'_{\gamma(i)} = \frac{\exp\left(\mathbf{X}^*_{\gamma(i)}\beta_\gamma\right)}{1+\exp\left(\mathbf{X}^*_{\gamma(i)}\beta_\gamma\right)}$.
- $f(\mathbf{y}^*|\beta_0, \delta) = \prod_{i=1}^{n^*}\left[p_0^{*\,y_i^*}(1-p_0^*)^{1-y_i^*}\right]$, where $p_0^* = \frac{\left(p'_0\right)^{1/\delta}}{\left(p'_0\right)^{1/\delta}+\left(1-p'_0\right)^{1/\delta}}$ and $p'_0 = \frac{\exp\beta_0}{1+\exp\beta_0}$.
- $\pi^{\mathrm{N}}(\beta_\gamma|\gamma) \propto 1$ and $\pi^{\mathrm{N}}(\beta_0) \propto 1$ (assuming reference baseline priors).

- $\pi^{\mathrm{N}}(\beta_{\setminus\gamma}|\gamma) = \mathrm{N}_{d_{\setminus\gamma}}\left(\widehat{\beta}_{[\gamma=0]}, \mathbf{I}_{d_{\setminus\gamma}}\widehat{\sigma_{\beta}}^{2}_{[\gamma=0]}\right)$, i.e. a multivariate normal distribution of dimensionality $d_{\setminus\gamma}$, where $\widehat{\beta}$ and $\widehat{\sigma_{\beta}}$ are the ML estimates and the corresponding standard errors, respectively, from the full model regressed on $\mathbf{y}$.

- $m_{\gamma}^{\mathrm{N}}(\mathbf{y}^*|\delta)$ is the prior predictive of model M_{γ} which for $\pi^{\mathrm{N}}(\beta_{\gamma}|\gamma) \propto 1$ is given by $m_{\gamma}^{\mathrm{N}}(\mathbf{y}^*|\delta) = \int f(\mathbf{y}^*|\beta_{\gamma}, \gamma, \delta)\mathrm{d}\beta_{\gamma}$. This density is estimated through the Laplace approximation as

$$\widehat{m}_{\gamma}^{\mathrm{N}}(\mathbf{y}^*|\tilde{\beta}_{\gamma}, \delta) = (2\pi)^{d_{\gamma}/2}|\tilde{\Sigma}|^{1/2} f(\mathbf{y}^*|\tilde{\beta}_{\gamma}, \gamma, \delta),$$

  where $\tilde{\beta}_{\gamma}$ is the posterior mode and $\tilde{\Sigma}$ is minus the inverse Hessian matrix evaluated at $\tilde{\beta}_{\gamma}$.

- $\pi^{\mathrm{N}}(\gamma) = \frac{1}{d+1}\binom{d}{d_{\gamma}}^{-1}$ for an appropriate multiplicity adjustment.

As recommended in Sect. 6.2.4, we set $\delta = n^*$ and $n^* = n$, for which we have $\mathbf{X}_{\gamma}^* \equiv \mathbf{X}_{\gamma}$. Given these specifications we implement GVS based on the following Metropolis-within-Gibbs sampling scheme:

A. Set starting values $\gamma^{(0)}, \beta_{\gamma}^{(0)}, \beta_0^{(0)}$ and $\mathbf{y}^{*(0)}$.

B. For iterations $t = 1, 2, \ldots, N$:

   1) Sampling of $\beta_{\gamma}^{(t)}$

      a) Given the state of γ and $\mathbf{y}^*$ at iteration $t-1$, generate $\beta_{\gamma}^{'}$ from the proposal distribution $q(\beta_{\gamma}) = \mathrm{N}_{d_{\gamma}}(\widehat{\beta}_{\gamma}, \widehat{\Sigma}_{\beta_{\gamma}})$, where $\widehat{\beta}_{\gamma}$ is the ML estimate from the regression on $\mathbf{y}^{\mathrm{all}} = (\mathbf{y}, \mathbf{y}^*)^T$, using weights $\mathbf{w}^{\mathrm{all}} = (\mathbf{1}_n, \mathbf{1}_n n^{-1})^T$, and $\widehat{\Sigma}_{\beta_{\gamma}}$ is the estimated variance–covariance matrix of $\widehat{\beta}_{\gamma}$.

      b) Calculate the probability of move:

$$\alpha_{\beta_{\gamma}} = \min\left[1, \frac{f(\mathbf{y}|\beta_{\gamma}^{'}, \gamma) f(\mathbf{y}^*|\beta_{\gamma}^{'}, \gamma, \delta) q(\beta_{\gamma}^{(t-1)})}{f(\mathbf{y}|\beta_{\gamma}^{(t-1)}, \gamma) f(\mathbf{y}^*|\beta_{\gamma}^{(t-1)}, \gamma, \delta) q(\beta_{\gamma}^{'})}\right].$$

      c) Set

$$\beta_{\gamma}^{(t)} = \begin{cases} \beta_{\gamma}^{'} & \text{with probability } \alpha_{\beta_{\gamma}}, \\ \beta_{\gamma}^{(t-1)} & \text{with probability } 1 - \alpha_{\beta_{\gamma}}. \end{cases}$$

2) Sampling of $\beta_{\setminus\gamma}^{(t)}$

a) Given the current state of γ at iteration $t-1$, generate $\beta_{\setminus\gamma}'$ from the pseudo-prior $\pi^{\mathrm{N}}(\beta_{\setminus\gamma}|\gamma) = \mathrm{N}_{d_{\setminus\gamma}}\left(\widehat{\beta}_{[\gamma=0]}, \mathbf{I}_{d_{\setminus\gamma}}\widehat{\sigma_{\beta}}^2_{[\gamma=0]}\right)$.

b) Set $\beta_{\setminus\gamma}^{(t)} = \beta_{\setminus\gamma}'$ with probability equal to 1.

3) Sampling of $\beta_0^{(t)}$

a) Given the state of $\mathbf{y}^*$ at iteration $t-1$, generate β_0' from the proposal distribution $q(\beta_0) = \mathrm{N}(\widehat{\beta_0}, \widehat{\sigma}^2_{\beta_0})$, where $\widehat{\beta_0}$ is the ML estimate from the regression on $\mathbf{y}^*$, using weights $\mathbf{w}^* = (\mathbf{1}_n n^{-1})^T$, and $\widehat{\sigma}_{\beta_0}$ is the standard error of $\widehat{\beta_0}$.

b) Calculate the probability of move:

$$\alpha_{\beta_0} = \min\left[1, \frac{f(\mathbf{y}^*|\beta_0', \delta)q(\beta_0^{(t-1)})}{f(\mathbf{y}^*|\beta_0^{(t-1)}, \delta)q(\beta_0')}\right].$$

c) Set

$$\beta_0^{(t)} = \begin{cases} \beta_0' & \text{with probability } \alpha_{\beta_0}, \\ \beta_0^{(t-1)} & \text{with probability } 1-\alpha_{\beta_0}. \end{cases}$$

4) Sampling of $\mathbf{y}^{*(t)}$

a) Given the current state of β_γ and β_0 at iteration t, generate $\mathbf{y}^{*'}$ from the proposal distribution $q(\mathbf{y}^*) = Bernoulli(p_{\mathbf{y}*})$, with the probability of success given by $p_{\mathbf{y}^*} = \frac{(p_{\beta_\gamma} p_{\beta_0})^{1/\delta}}{(p_{\beta_\gamma} p_{\beta_0})^{1/\delta} + (1-p_{\beta_\gamma} p_{\beta_0})^{1/\delta}}$, where $p_{\beta_\gamma} = \frac{\exp(\mathbf{X}_\gamma\beta_\gamma)}{1+\exp(\mathbf{X}_\gamma\beta_\gamma)}$ and $p_{\beta_0} = \frac{\exp\beta_0}{1+\exp\beta_0}$.

b) Calculate the probability of move:

$$\alpha_{\mathbf{y}^*} = \min\left[1, \frac{f(\mathbf{y}^{*'}|\beta_\gamma, \gamma, \delta) f(\mathbf{y}^{*'}|\beta_0, \delta)\widehat{m}_\gamma^{\mathrm{N}}(\mathbf{y}^{*(t-1)}|\tilde{\beta}_\gamma, \delta)q(\mathbf{y}^{*(t-1)})}{f(\mathbf{y}^{*(t-1)}|\beta_\gamma, \gamma, \delta) f(\mathbf{y}^{*(t-1)}|\beta_0, \delta)\widehat{m}_\gamma^{\mathrm{N}}(\mathbf{y}^{*'}|\tilde{\beta}_\gamma, \delta)q(\mathbf{y}^{*'})}\right].$$

c) Set

$$\mathbf{y}^{*(t)} = \begin{cases} \mathbf{y}^{*'} & \text{with probability } \alpha_{\mathbf{y}^*}, \\ \mathbf{y}^{*(t-1)} & \text{with probability } 1-\alpha_{\mathbf{y}^*}. \end{cases}$$

5) Sampling of $\gamma_j^{(t)}$ for $j = 1, 2, \ldots, d$

   a) For the current state of $\beta_\gamma, \beta_{\setminus\gamma}$ and $\mathbf{y}^*$, calculate the current odds

$$\mathrm{CO}(\gamma_j) = \frac{\dfrac{f(\mathbf{y}|\beta_\gamma,\gamma_j=1,\gamma_{\setminus j})f(\mathbf{y}^*|\beta_\gamma,\gamma_j=1,\gamma_{\setminus j},\delta)\pi^{\mathrm{N}}(\beta_{\setminus\gamma}|\gamma_j=1,\gamma_{\setminus j})}{\widehat{m}_\gamma^{\mathrm{N}}(\mathbf{y}^*|\tilde{\beta}_\gamma,\gamma_j=1,\gamma_{\setminus j},\delta)}}{\dfrac{f(\mathbf{y}|\beta_\gamma,\gamma_j=0,\gamma_{\setminus j})f(\mathbf{y}^*|\beta_\gamma,\gamma_j=0,\gamma_{\setminus j},\delta)\pi^{\mathrm{N}}(\beta_{\setminus\gamma}|\gamma_j=0,\gamma_{\setminus j})}{\widehat{m}_\gamma^{\mathrm{N}}(\mathbf{y}^*|\tilde{\beta}_\gamma,\gamma_j=0,\gamma_{\setminus j},\delta)}}$$

   and the prior odds

$$\mathrm{PrO}(\gamma_j) = \frac{d_j + 1}{d - d_j},$$

   where $d_j = \sum_{i \neq j} \gamma_i$.

   b) Calculate $\mathrm{O}(\gamma_j) = \mathrm{CO}(\gamma_j) \times \mathrm{PrO}(\gamma_j)$

   c) Sample $\gamma_j' \sim \text{Bernoulli}\left(\frac{\mathrm{O}(\gamma_j)}{1+\mathrm{O}(\gamma_j)}\right)$ and set $\gamma_j^{(t)} = \gamma_j'$ with probability equal to 1.

C. Repeat the steps in B until convergence.

References

[1] Bartlett, M.S.: Comment on D.V. Lindley's statistical paradox. Biometrika **44**, 533–534 (1957)

[2] Berger, J.O., Pericchi, L.R.: The intrinsic Bayes factor for model selection and prediction. J. Am. Stat. Assoc. **91**, 109–122 (1996)

[3] Casella, G., Moreno, E.: Objective Bayesian variable selection. J. Am. Stat. Assoc. **101**, 157–167 (2006)

[4] Dellaportas, P., Forster, J.J., Ntzoufras, I.: On Bayesian model and variable selection using MCMC. Stat. Comput. **12**, 27–36 (2002)

[5] Fernández, C., Ley, E., Steel, M.F.J.: Benchmark priors for Bayesian model averaging. J. Econ. **100**, 381–427 (2001)

[6] Fouskakis, D., Ntzoufras, I.: Computation for intrinsic variable selection in normal regression models via expected-posterior prior. Stat. Comput. **23**, 491–499 (2013)

[7] Fouskakis, D., Ntzoufras, I.: Power-conditional-expected priors: using g-priors with random imaginary data for variable selection. J. Comput. Graph. Statist. Forthcoming (2015)

[8] Fouskakis, D., Ntzoufras, I., Draper, D.: Power-expected-posterior priors for variable selection in Gaussian linear models. Bayesian Anal. **10**, 75–107 (2015)

[9] Holmes, C.C., Held, L.: Bayesian auxiliary variable models for binary and multinomial regression. Bayesian Anal. **1**, 145–168 (2006)

[10] Ibrahim, J., Chen, M.: Power prior distributions for regression models. Stat. Sci. **15**, 46–60 (2000)

[11] Kass, R., Wasserman, L.: A reference Bayesian test for nested hypotheses and its relationship to the Schwarz criterion. J. Am. Stat. Assoc. **90**, 928–934 (1995)

[12] Ley, E., Steel, M.F.J.: Mixtures of g-priors for Bayesian model averaging with economic applications. J. Econ. **171**, 251–266 (2012)

[13] Liang, F., Paulo, R., Molina, G., Clyde, M.A., Berger, J.O.: Mixtures of g-priors for Bayesian variable selection. J. Am. Stat. Assoc. **103**, 410–423 (2008)
[14] Lindley, D.V.: A statistical paradox. Biometrika **44**, 187–192 (1957)
[15] Murray, I., Ghahramani, Z., MacKay, D.J.C.: MCMC for doubly-intractable distributions. In: Proceedings of the 22nd Annual Conference on Uncertainty in Artificial Intelligence (UAI-06), pp. 359–366. AUAI Press, Arlington (2006)
[16] Ntzoufras, I., Dellaportas, P., Forster, J. J.: Bayesian variable and link determination for generalised linear models. Stat. Plann. Inference **111**, 165–180 (2003)
[17] O'Hagan, A.: Fractional Bayes factors for model comparison. J. R. Stat. Soc. Ser. B Stat. Methodol. **57**, 99–138 (1995)
[18] Pérez, J.M., Berger, J.O.: Expected-posterior prior distributions for model selection. Biometrika **89**, 491–511 (2002)
[19] Ripley, B.D.: Pattern Recognition and Neural Networks. Cambridge University Press, Cambridge (1996)
[20] Sabanés Bové, D., Held, L.: Hyper-g priors for generalized linear models. Bayesian Anal. **6**, 387–410 (2011)
[21] Scott, J.G., Berger, J.O.: Bayes and empirical-Bayes multiplicity adjustment in the variable-selection problem. Ann. Stat. **38**, 2587–2619 (2010)
[22] Spiegelhalter, D., Abrams, K., Myles, J.: Bayesian Approaches to Clinical Trials and Health-Care Evaluation, Statistics in Practice. Wiley, Chichester (2004)
[23] Tüchler, R.: Bayesian variable selection for logistic models using auxiliary mixture sampling. J. Comput. Graph. Stat. **17**, 76–94 (2008)
[24] Zellner, A.: On assessing prior distributions and Bayesian regression analysis using g-prior distributions. In: Goel, P., Zellner, A. (eds.) Bayesian Inference and Decision Techniques: Essays in Honor of Bruno de Finetti, pp. 233–243. North-Holland, Amsterdam (1986)
[25] Zellner, A., Siow, A.: Posterior odds ratios for selected regression hypotheses. In: Bernardo, J.M., DeGroot, M.H., Lindley, D.V., Smith, A.F.M. (eds.) Bayesian Statistics: Proceedings of the First International Meeting Held in Valencia, pp. 388–396. University of Valencia Press, Valencia (1980)

Chapter 7
Application of Interweaving in DLMs to an Exchange and Specialization Experiment

Matthew Simpson

Abstract Markov chain Monte Carlo is often particularly challenging in dynamic models. In state space models, the data augmentation algorithm (Tanner and Hung Wong, J. Am. Stat. Assoc. 82(398):528–540, 1987) is a commonly used approach, e.g. (Carter and Kohn, Biometrika 81(3):541–553, 1994) and (Frühwirth-Schnatter, J. Time Ser. Anal. 15(2):183–202, 1994) in dynamic linear models. Using two data augmentations, Yu and Meng (J. Comput. Graph. Stat. 20(3): 531–570, 2011) introduces a method of "interweaving" between the two augmentations in order to construct an improved algorithm. Picking up on this, Simpson et al. (Interweaving Markov chain Monte Carlo strategies for efficient estimation of dynamic linear models, Working Paper, 2014) introduces several new augmentations for the dynamic linear model and builds interweaving algorithms based on these augmentations. In the context of a multivariate model using data from an economic experiment intended to study the disequilibrium dynamics of economic efficiency under a variety of conditions, we use these interweaving ideas and show how to implement them simply despite complications that arise because the model has latent states with a higher dimension than the data.

Key words: Ancillary augmentation, Centered parameterization, Data augmentation, Non-centered parameterization, Reparameterization, Sufficient augmentation, Time series, State-space model

7.1 Introduction

Several innovations on the original data augmentation (DA) algorithm [25] have been proposed in the literature, see, e.g., [26] for a thorough overview. One such innovation is the notion of interweaving two separate DAs together [27].

M. Simpson (✉)
Department of Economics, Iowa State University, Ames, IA, USA
e-mail: themattsimpson@gmail.com

S. Frühwirth-Schnatter et al. (eds.), *Bayesian Statistics from Methods to Models and Applications*, Springer Proceedings in Mathematics & Statistics 126,
DOI 10.1007/978-3-319-16238-6_7

This general idea has been picked up on in the dynamic setting by [15] in stochastic volatility models and [23] in dynamic linear models (DLMs). Previous literature exploring alternate DAs in state space models includes [19] for the AR(1) plus noise model, [6] for dynamic regression models, [24] for nonlinear models including the stochastic volatility model, [8] for the stochastic volatility model, and [9] in the context of model selection, though there are many more.

Much of this literature focuses on stochastic volatility and similar models [1, 7, 8, 15, 20, 22, 24], though [23] focuses on DLMs and develops several new data augmentations for a general class of DLMs. Using these DAs, they construct several Markov chain Monte Carlo (MCMC) algorithms including interweaving algorithms based on [27], and compare these algorithms in a simulation study using the local level model. We seek to illustrate the interweaving methods introduced in [23] in the context of models that can be expressed either as a hierarchical DLM with equal state and data dimensions or simply a DLM with a state dimension larger than the data dimension. The latter representation in particular provides some difficulty in directly applying the methods discussed in [23], though we show how to easily overcome this.

Throughout this article we will use the notation $p(.|.)$ to denote the potentially conditional density of the enclosed random variables, $x_{1:T} = (x_1, \ldots, x_T)'$ when x_t is a scalar, and $x_{1:T} = (x_1', \ldots, x_T')'$ when x_t is a column vector so that $x_{1:T}$ is also a column vector in both cases. The rest of this paper is organized as follows: Sect. 7.2 will describe the data which arise from a series of economic experiments, and Sect. 7.3 will describe the model we wish to fit to these data. Section 7.4 will cover how to do MCMC in this model, including a fairly standard DA algorithm and an interweaving algorithm based on the ideas in [23] and [27]. Finally, Sect. 7.5 will contain the results of fitting the model using both algorithms, and Sect. 7.6 will briefly conclude.

7.2 Data

Economists are interested in determining the factors that affect the level of economic efficiency within an economy where economic efficiency can roughly be defined as the proportion of maximum possible dollar value of the total benefits to all actors in the economy, also known as Kaldor–Hicks efficiency and based on compensating variation [14, 17]. Studying this in the real world is messy and difficult in part because computing this proportion is nontrivial. In addition, most economic models only allow the analysis of equilibrium efficiency. To the extent that efficiency dynamics are studied, they are typically studied as equilibrium dynamics. Disequilibrium dynamics are difficult to study but potentially important. In order to avoid these difficulties while still learning something about the disequilibrium dynamics of efficiency, a series of laboratory experiments were designed and run by a group of experimental economists in order to explore what factors impact the disequilibrium dynamics of a small laboratory economy [4, 16]. What follows is a brief description of these experiments.[1]

In a single session of the experiment, 2, 4, or 8 subjects are recruited to participate, depending on the treatment. Each subject sits at a computer visually isolated from the rest of the subjects. On the computer, each subject controls an avatar in a virtual village where they can interact with the other subjects in the experiment. At any time during the experiment, subjects can communicate with each other by typing into a chat window. Each subject in a given session has control over a house and a field within the village and can view each other subject's house and field. The experiment runs for 40 periods, each lasting 100 s. Within a period, each subject has to make a production decision and a consumption decision. Every seventh period is a "rest" period where no production or consumption takes place, but the subjects can still communicate. This results in 35 periods of production and consumption.

There are two types of goods in this world, each produced in a subject's field: *red* and *blue*, and two types of subjects: *odd* and *even*. Half of the subjects are *odd* and half are *even*. Both *odd* and *even* subjects can produce both types of goods and earn money for consuming both types of goods, but they produce and consume in different ways. *Odd* subjects must consume *red* and *blue* in a fixed proportion of one *red* for every three *blue* to earn U.S. cents. *Even* subjects, on the other hand, must consume two *red* for every one *blue* to earn U.S. cents. However, *even* subjects are more effective at producing *blue* while *odd* subjects are more effective at producing *red*. Production occurs in the first 10 s of a period where each subject must decide how much of that time to devote to producing *red* and *blue*, respectively using a slider on their screen. The last 90 s of the period is reserved for trading and consumption, though subjects have to discover that they may trade by noticing that they can use their mouse to drag and drop red and/or blue icons (representing one unit of *red* or *blue*, respectively) onto another subject's house. The maximum level of village wide production takes place when each subject spends 100 % of their time producing the good that they can produce the most efficiently, i.e. *odd* subjects produce only *red* and *even* subjects produce only *blue*. Maximum consumption and thus maximum profit occurs when under maximum production and the subjects trade extensively with each other. In every period, the efficiency level of the village is recorded.

A wide variety of treatments were applied to the various sessions of this experiment, including variations on group size and group formation, various levels of knowledge about the subject's own production function, allowing theft or not and if so, whether mechanisms for punishing theft are available. See [4] and [16] for a detailed description of these treatments. Each treatment consists of several replications—anywhere from four to six. The challenge, then, is to model a time series of proportions that takes into account the nested structure of the replications

[1]For a more detailed description of the experimental design, see [4] especially, but also [16].

within the treatments. To deal with the proportions, we simply transform the efficiencies to the real line using the logit transformation, i.e. $\text{logit}(x) = \log(x/(1-x))$. In some replications of some treatments, efficiencies of 100 % or 0 % are obtained which causes a problem for the logit and other plausible transformations. We only consider the Steal treatment of [16] in order to avoid this issue and simplify the model a bit. This allows for a useful illustration of [23] without too much additional complication. In short, the Steal treatment uses the Build8 structure from previous treatments that starts the subjects in four groups of two for several periods, then combines them into two groups of four for several more periods, then finally combines the groups into a single group of eight for the rest of the experiment. The only change from this structure is that Steal allows subjects to steal either of the goods from each other, which was not possible in previous treatments. Kimbrough et al. [16] has further details about this treatment and the various treatments it spawned in order to see what institutional arrangements help subjects prevent theft.

7.3 Model

Let $j = 1, 2, \ldots, J$ denote the replications of the treatment and $t = 1, 2, \ldots, T$ denote periods within these replications. Then let $y_{j,t}$ denote the observed logit efficiency of the jth replication in the tth period. Consider the following model

$$\begin{aligned} y_{j,t} &= \mu_t + \theta_{j,t} + v_{j,t} && \text{(observation equation)} \\ \theta_{j,t} &= \theta_{j,t-1} + w_{j,t} && \text{(replication level system equation)} \\ \mu_t &= \mu_{t-1} + u_t && \text{(treatment level system equation)} \end{aligned} \tag{7.1}$$

for $j = 1, 2, \ldots, J$ and $t = 1, 2, \ldots, T$, where $(v_{1:J,1:T}, w_{1:J,1:T}, u_{1:T})$ are mutually independent with $v_{j,t} \sim N(0, V_j)$, $w_{j,t} \sim N(0, W_j)$, and $u_t \sim N(0, U)$. The latent treatment level logit efficiency is represented by μ_t and evolves via a random walk. On the replication level, $\theta_{j,t}$ represents replication j's deviation from the treatment logit efficiency in period t which also evolves over time via a random walk. Then $\mu_t + \theta_{j,t}$ is replication level latent logit efficiency. Finally $y_{j,t}$ represents the observed logit efficiency of replication j in period t. The amount replication jth path tends to differ from the treatment level path is controlled by the relative values of W_j and U—the larger W_j is relative to U, the less replication jth path is affected by the treatment level path. Finally, V_j represents how much of the change in logit efficiency is independent of previous changes. The relative size of V_j compared to W_j and U tells us how much logit efficiency changes over time due to independent sources of error relative to the replication and treatment level evolutionary processes. So in this sense, $\mu_t + \theta_{j,t}$ can be seen as the portion of replication j's logit efficiency that is carried on into the next period, or sustainable in a certain sense.

Another way to represent this model is by writing it in terms of the replication level latent logit efficiencies, $\phi_{j,t} = \mu_t + \theta_{j,t}$. Under this parameterization, the model is

$$\begin{aligned} y_{j,t} &= \phi_{j,t} + v_{j,t}, \\ \phi_{j,t} &= \phi_{j,t-1} + w_{j,t} + u_t, \end{aligned} \tag{7.2}$$

for $j = 1,2,\ldots,J$ and $t = 1,2,\ldots,T$, where we substitute u_t for $\mu_t - \mu_{t-1}$. This representation shows us that the replication level latent logit efficiencies evolve according to a correlated random walk where U controls the degree of correlation between the replications.

Finally, if we let $\theta_t = (\mu_t, \theta'_{1:J,t})'$, $y_t = y_{1:J,t}$, $V = diag(V_1,\ldots,V_J)$, $W = diag(U, W_1,\ldots,W_J)$, and $F = [1_{J\times 1}\ \ I_{J\times J}]$, we can write the model as a multivariate DLM:

$$\begin{aligned} y_t|\theta_{0:T} &\sim N_J(F\theta_t, V) \\ \theta_t|\theta_{0:(t-1)} &\sim N_{J+1}(\theta_{t-1}, W) \end{aligned} \tag{7.3}$$

for $t = 1,2,\ldots,T$. This representation will be useful for constructing MCMC algorithms for the model. Using this representation, we need priors for the V_j's, W_j's, U, and θ_0 to complete the model. We will suppose that they are independent with $\theta_0 \sim N_{J+1}(m_0, C_0)$, $V_j \sim IG(a_{V_j}, b_{V_j})$, $W_j \sim IG(a_{W_j}, b_{W_j})$, and $U \sim IG(a_U, b_U)$. We will set $m_0 = 0_{J+1}$, $C_0 = diag(100)$, $a_{V_j} = a_{W_j} = a_u = 1.5$ and $b_{V_j} = b_{W_j} = b_U = 0.25$. This prior on the variance parameters has essentially zero mass below 0.02 and above 2, which allows for a fairly wide range of parameter estimates relative to the scale of the data. These priors are chosen for convenience in illustrating the MCMC method of [23] and for simplicity, but a simple way to use the inverse-gamma priors without their well-known inferential problems [10] is to put gamma hyperpriors on the b parameters rather than fixing them. The marginal priors on the standard deviations will then be half-t and in the MCMC samplers we discuss a Gibbs step will have to be added for drawing the b's from a gamma distribution. This prior is the hierarchical inverse-Wishart prior of [13] in the scalar case.

7.4 Markov Chain Monte Carlo

We construct two separate MCMC samplers for this model. One is a naive data augmentation algorithm and the other takes advantage of the interweaving technology of [27], particularly the developments of [23] for DLMs. We primarily use the DLM representation of the model given in (7.3).

7.4.1 Naive Data Augmentation

The standard DA algorithm characterizes the posterior of (V,W) by using a Gibbs sampler to draw from the posterior distribution of $(V,W,\theta_{0:T})$ [25]. In this particular case we are also interested in the posterior of $\theta_{0:T}$, which is common in dynamic models, but this does not change the MCMC strategy. The sampler is based on [5] and [3] and consists of two steps, a draw from $p(\theta_{0:T}|V,W,y_{1:T})$ and a draw from $p(V,W|\theta_{0:T},y_{1:T})$. In order to construct this algorithm we need these two densities.

First, from the DLM representation of the model in (7.3), and the priors we can write the joint posterior density of V, W, and $\theta_{0:T}$ as

$$\begin{aligned}
p(V,W,\theta_{0:T}|y_{1:T}) \propto {} & |V|^{-T/2}\exp\left[-\frac{1}{2}\sum_{t=1}^{T}(y_t - F\theta_t)'V^{-1}(y_t - F\theta_t)\right] \\
& \times |W|^{-T/2}\exp\left[-\frac{1}{2}\sum_{t=1}^{T}(\theta_t - \theta_{t-1})'W^{-1}(\theta_t - \theta_{t-1})'\right] \\
& \times \exp\left[-\frac{1}{2}(\theta_0 - m_0)'C_0^{-1}(\theta_0 - m_0)\right] U^{-a_U-1}\exp\left[-\frac{1}{U}b_U\right] \\
& \times \prod_{j=1}^{J} V_j^{-a_{V_j}-1}\exp\left[-\frac{1}{V_j}b_{V_j}\right] W_j^{-a_{W_j}-1}\exp\left[-\frac{1}{W_j}b_{W_j}\right].
\end{aligned} \tag{7.4}$$

From here we can derive the smoothing density, or conditional posterior density of $\theta_{0:T}$. We use the method of [18], based on [21], for drawing from this density, called the mixed Cholesky factor algorithm (MCFA) by [23]. The following derivation closely follows Appendix C of [23]. The full conditional density of $\theta_{0:T}$ can be written as

$$p(\theta_{0:T}|V,W,y_{1:T}) \propto \exp\left[-\frac{1}{2}g(\theta_{0:T})\right]$$

where

$$\begin{aligned}
g(\theta_{0:T}) = {} & \sum_{t=1}^{T}(y_t - F\theta_t)'V^{-1}(y_t - F\theta_t) + \sum_{t=1}^{T}(\theta_t - \theta_{t-1})'W^{-1}(\theta_t - \theta_{t-1}) \\
& + (\theta_0 - m_0)'C_0^{-1}(\theta_0 - m_0).
\end{aligned}$$

Then g has the form $g(\theta_{0:T}) = \theta_{0:T}'\Omega\theta_{0:T} - 2\theta_{0:T}'\omega + K$ where K is some constant with respect to $\theta_{0:T}$, Ω is a square, symmetric matrix of dimension $(J+1)(T+1)$ and ω is a column vector of dimension $(J+1)(T+1)$. This gives $\theta_{0:T}|V,W,y_{1:T} \sim N_{(J+1)(T+1)}(\Omega^{-1}\omega,\Omega^{-1})$. Further, Ω is block tridiagonal since there are no cross product terms involving θ_t and θ_{t+k} where $|k|>1$. Because of this, the Cholesky factor and thus inverse of Ω can be efficiently computed leading to the Cholesky factor algorithm (CFA) [21]. Instead of computing the Cholesky factor of Ω all at once before drawing $\theta_{0:T}$ as in the CFA, the same technology can be used to draw θ_T, then $\theta_t|\theta_{(t+1):T}$ recursively in a backward sampling structure, resulting in the MCFA. In simulations, the MCFA has been found to be significantly cheaper than Kalman filter based methods and often cheaper than the CFA [18].

In order to implement the algorithm, we need to first characterize the diagonal and off diagonal blocks of Ω and the blocks of ω:

$$\begin{aligned}
\Omega_{0,0} &= C_0^{-1} + G_1' W^{-1} G_1 \\
\Omega_{t,t} &= F'V^{-1}F + 2W^{-1} && \text{for } t = 1,2,\ldots T-1 \\
\Omega_{T,T} &= F'V^{-1}F + W^{-1} \\
\Omega_{t,t-1} &= -W_t^{-1} = \Omega_{t-1,t} && \text{for } t = 1,2,\ldots T \\
w_0 &= C_0^{-1} m_0 \\
w_t &= F'V^{-1} y_t && \text{for } t = 1,2,\ldots T.
\end{aligned}$$

Now let $\Sigma_0 = \Omega_{0,0}^{-1}$, $\Sigma_t = (\Omega_{t,t} - \Omega_{t,t-1}\Sigma_{t-1}\Omega_{t-1,t})^{-1}$ for $t = 1,2,\ldots,T$, $h_0 = \Sigma_0 w_0$, and $h_t = \Sigma_t(w_t - \Omega_{t,t-1}h_{t-1})$ for $t = 1,2,\ldots,T$. Then to complete the MCFA we perform the following draws recursively

$$\begin{aligned}
\theta_T &\sim N(h_T, \Sigma_T) \\
\theta_t|\theta_{(t+1):T} &\sim N(h_t - \Sigma_t\Omega_{t,t+1}\theta_{t+1}, \Sigma_t) && \text{for } t = T-1, T-2, \ldots, 0.
\end{aligned}$$

The second step of the DA algorithm requires a draw from $p(V,W|\theta_{0:T}, y_{1:T})$. Recalling that $V = diag(V_1,\ldots,V_J)$ and $W = diag(U, W_1,\ldots,W_J)$, this density is

$$\begin{aligned}
p(V,W|\theta_{0:T}, y_{1:T}) \propto\; & U^{-a_U - T/2 - 1} \exp\left[-\frac{1}{U}\left(b_U + \frac{1}{2}\sum_{t=1}^{T}(\mu_t - \mu_{t-1})^2\right)\right] \\
&\times \prod_{j=1}^{J} V_j^{-a_{V_j} - T/2 - 1} \exp\left[-\frac{1}{V_j}\left(b_{V_j} + \frac{1}{2}\sum_{t=1}^{T}(y_{j,t} - \mu_t - \theta_{j,t})^2\right)\right] \\
&\times \prod_{j=1}^{J} W_j^{-a_{W_j} - T/2 - 1} \exp\left[-\frac{1}{W_j}\left(b_{W_j} + \frac{1}{2}\sum_{t=1}^{T}(\theta_{j,t} - \theta_{j,t-1})^2\right)\right].
\end{aligned}$$

This is the product of inverse-gamma densities, so a draw from this density can easily be accomplished by

$$\begin{aligned}
V_j &\sim IG(\tilde{a}_{V_j}, \tilde{b}_{V_j}) && \text{for } j = 1,2,\ldots,J \\
W_j &\sim IG(\tilde{a}_{W_j}, \tilde{b}_{W_j}) && \text{for } j = 1,2,\ldots,J \\
U &\sim IG(\tilde{a}_U, \tilde{b}_U)
\end{aligned}$$

where $\tilde{a}_U = a_U + T/2$, $\tilde{b}_U = b_U + \sum_{t=1}^{T}(\mu_t - \mu_{t-1})^2/2$, and for $j = 1,2,\ldots,J$, $\tilde{a}_{V_j} = a_{V_j} + T/2$, $\tilde{b}_{V_j} = b_{V_j} + \sum_{t=1}^{T}(y_{j,t} - \mu_t - \theta_{j,t})^2/2$, $\tilde{a}_{W_j} = a_{W_j} + T/2$, and $\tilde{b}_{W_j} = b_{W_j} + \sum_{t=1}^{T}(\theta_{j,t} - \theta_{j,t-1})^2/2$. So we can write the naive DA algorithm as follows:

1. Draw $\theta_{0:T} \sim N(\Omega^{-1}\omega, \Omega^{-1})$ using the MCFA.
2. Draw $U \sim IG(\tilde{a}_U, \tilde{b}_U)$.
3. For $j = 1,2,\ldots,J$ draw $V_j \sim IG(\tilde{a}_{V_j}, \tilde{b}_{V_j})$ and $W_j \sim IG(\tilde{a}_{W_j}, \tilde{b}_{W_j})$.

Note that step 2 and the $2J$ sub-steps of step 3 can be parallelized since the draws are all independent, though we do not explore this possibility.

7.4.2 Interweaving

The basic idea of interweaving is to use two separate DAs and "weave" them together [27]. Suppose we have the DAs $\gamma_{0:T}$ and $\psi_{0:T}$. Then an alternating algorithm for our model consists of four steps:

$$[\gamma_{0:T}|V,W,y_{1:T}] \to [V,W|\gamma_{0:T},y_{1:T}] \to [\psi_{0:T}|V,W,y_{1:T}] \to [V,W|\psi_{0:T},y_{1:T}].$$

The first two steps are simply the two steps of the DA algorithm based on $\gamma_{0:T}$ while the last two steps are the two steps of the DA algorithm based on $\psi_{0:T}$. A global interweaving strategy (GIS) using these two augmentations is very similar:

$$[\gamma_{0:T}|V,W,y_{1:T}] \to [V,W|\gamma_{0:T},y_{1:T}] \to [\psi_{0:T}|V,W,\gamma_{0:T},y_{1:T}] \to [V,W|\psi_{0:T},y_{1:T}].$$

The only difference is that in step 3, we condition on $\gamma_{0:T}$ as well as V, W, and $y_{1:T}$. Often, this is a transformation using the definition of $\gamma_{0:T}$ and $\psi_{0:T}$, and not a random draw. When step 3 is a transformation, this reduces the computational cost relative to the alternating algorithm. Depending on the properties of the data augmentations used, changing step 3 in this manner can also drastically improve the behavior of the Markov chain whether or not step 3 is a transformation [27].

Simpson et al. [23] defines several DAs for the DLM, including the following two—the scaled disturbances, defined by $\gamma_t = L_W^{-1}(\theta_t - \theta_{t-1})$, and the scaled errors, defined by $\psi_t = L_V^{-1}(y_t - F\theta_t)$ for $t = 1,2,\ldots,T$ and $\psi_0 = \gamma_0 = \theta_0$ where L_X denotes the lower triangular Cholesky factor of the symmetric and positive definite matrix X. Since the dimension of y_t and θ_t is not the same, the scaled errors cannot be directly used without some additional augmentation. Another option is to use a representation of the model which removes the treatment level states, given in (7.2). Using this is unwieldy because the full conditional posterior of $(W_{1:J}, U)$ becomes complicated since the $\phi_{j,t}$'s are correlated across groups. Instead of either of those, we will take a particularly simple approach. Consider the hierarchical representation of the model given in (7.1). For $j = 1,2,\ldots,J$ define the replication level scaled disturbances as $\gamma_{j,t} = (\theta_{j,t} - \theta_{j,t-1})/\sqrt{W_j}$ for $t = 1,2,\ldots,T$ and $\gamma_{j,0} = \theta_{j,0}$ and the replication level scaled errors as $\psi_{j,t} = (y_{j,t} - \mu_t - \theta_{j,t})/\sqrt{V_j}$ for $t = 1,2,\ldots,T$ and $\psi_{j,0} = \theta_{j,0}$. Now let $\gamma_t = (\mu_t, \gamma_{1:J,t}')'$ and $\psi_t = (\mu_t, \psi_{1:J,t}')'$ Then we can easily interweave between $\gamma_{0:T}$ and $\psi_{0:T}$ since these are one-to-one transformations of each other. Specifically the GIS algorithm we seek to construct is

1. Draw $\gamma_{0:T} \sim p(\gamma_{0:T}|V_{1:J}, W_{1:J}, U, y_{1:T})$.
2. Draw $(V_{1:J}, W_{1:J}, U) \sim p(V_{1:J}, W_{1:J}, U|\gamma_{0:T}, y_{1:T})$
3. Transform $\gamma_{0:T} \to \psi_{0:T}$ and draw $(V_{1:J}, W_{1:J}, U) \sim p(V_{1:J}, W_{1:J}, U|\psi_{0:T}, y_{1:T})$.

In order to complete this algorithm, we need to characterize the relevant full conditionals. First, consider the transformation from $\theta_{j,0:T}$ to $\gamma_{j,0:T}$. The Jacobian is triangular with a one and T copies of $\sqrt{W_j}$ along the diagonal. So the joint posterior of $V_{1:T}, W_{1:J}, U$, and $\gamma_{0:T}$ is

$$p(V_{1:T}, W_{1:J}, U, \gamma_{0:T}|y_{1:T}) \propto U^{-a_U - T/2 - 1} \exp\left[-\frac{1}{U}\left(b_U + \frac{1}{2}\sum_{t=1}^{T}(\mu_t - \mu_{t-1})^2\right)\right]$$

$$\times \exp\left[-\frac{1}{2}\sum_{j=1}^{J}\sum_{t=1}^{T}\gamma_{j,t}^2\right] \exp\left[-\frac{1}{2}(m_0 - \gamma_0)' C_0^{-1}(m_0 - \gamma_0)\right]$$

$$\times \prod_{j=1}^{J} V_j^{-a_{V_j} - T/2 - 1} \exp\left[-\frac{1}{V_j}\left(b_{V_j} + \frac{1}{2}\sum_{t=1}^{T}\left(y_{j,t} - \mu_t - \gamma_{j,0} - \sqrt{W_j}\sum_{s=1}^{t}\gamma_{j,s}\right)^2\right)\right]$$

$$\times \prod_{j=1}^{J} W_j^{-a_{W_j} - 1} \exp\left[-\frac{1}{W_j} b_{W_j}\right].$$

This allows us to write the model as

$$y_{j,t} = \mu_t + \sqrt{W_j}\sum_{s=1}^{t}\gamma_{j,s} + \gamma_{j,0} + v_{j,t},$$

$$\mu_t = \mu_{t-1} + u_t, \tag{7.5}$$

where $(v_{1:J,1:T}, \gamma_{1:J,1:T}, u_{1:T})$ are mutually independent with $\gamma_{j,t} \sim N(0,1)$, $v_{j,t} \sim N(0, V_j)$, and $u_t \sim N(0, U)$ for $j = 1,2,\ldots,J$ and $t = 1,2,\ldots,T$. The full conditional of $\gamma_{0:T}$ is a bit more complicated than that of $\theta_{0:T}$, but we can just use the MCFA to draw from $\theta_{0:T}$'s full conditional and transform to $\gamma_{0:T}$. The full conditional of $(V_{1:J}, W_{1:J}, U)$ is

$$p(V_{1:T}, W_{1:J}, U|\gamma_{0:T}, y_{1:T}) = p(U|\gamma_{0:T}, y_{1:T}) \prod_{j=1}^{J} p(V_j, W_j|\gamma_{0:T}, y_{1:T}).$$

Here, $p(U|\gamma_{0:T},y_{1:T}) = p(U|\theta_{0:T},y_{1:T})$, i.e. the same inverse-gamma distribution as when we conditioned on $\theta_{0:T}$. However, $p(V_j,W_j|\gamma_{0:T},y_{1:T})$ is complicated and difficult to sample from efficiently. Instead of drawing V_j and W_j jointly, we draw from their full conditionals. It turns out that $V_j|W_j,\gamma_{0:T},y_{1:T} \sim IG(\tilde{a}_{V_j},\tilde{b}_{V_j})$, which is the same as when we conditioned on $\theta_{0:T}$. The full conditional density is of W_j is still rather complicated:

$$p(W_j|V_j,\gamma_{0:T},y_{1:T}) \propto W_j^{-a_{W_j}-1} \exp\left[-b_{W_j}\frac{1}{W_j} + c_{W_j}\sqrt{W_j} - d_{W_j}W_j\right],$$

where

$$c_{W_j} = \frac{\sum_{t=1}^{T}(y_{j,t} - \mu_t - \gamma_{j,0})\sum_{s=1}^{t}\gamma_{j,s}}{V_j} \in \mathfrak{R}, \qquad d_{W_j} = \frac{\sum_{t=1}^{T}\left(\sum_{s=1}^{t}\gamma_{j,s}\right)^2}{2V_j} > 0.$$

The double summations in c_{W_j} and d_{W_j} are one consequence of the model no longer having the Markov property, which can easily be seen from (7.5). These summations can be expensive for large datasets, though in our experience this is typically not the most important computational bottleneck. In any case the summations can be attained much more efficiently via parallelization, especially using a GPU. In order to sample from this density, we follow [23] (Appendix E) and use an adaptive rejection sampling approach [12] when it is log concave, and otherwise we use a Cauchy approximation in a rejection sampling scheme for the density of $\log(W_j)$.

Now, we need to characterize the full conditionals given $\psi_{0:T}$. The Jacobian matrix of the transformation from $\theta_{j,0:T}$ to $\psi_{j,0:T}$ is diagonal with a one and T copies of $\sqrt{V_j}$ along the diagonal. So the joint posterior of $V_{1:T}, W_{1:J}, U$, and $\psi_{0:T}$ is

$$\begin{aligned}
p(V_{1:T},W_{1:J},U,\psi_{0:T}|y_{1:T}) &\propto U^{-a_U-T/2-1}\exp\left[-\frac{1}{U}\left(b_U + \frac{1}{2}\sum_{t=1}^{T}(\mu_t-\mu_{t-1})^2\right)\right] \\
&\times \exp\left[-\frac{1}{2}\sum_{j=1}^{J}\sum_{t=1}^{T}\psi_{j,t}^2\right]\exp\left[-\frac{1}{2}(m_0-\psi_0)'C_0^{-1}(m_0-\psi_0)\right] \\
&\times \prod_{j=1}^{J} W_j^{-a_{W_j}-T/2-1}\exp\left[-\frac{1}{W_j}\left(b_{W_j} + \frac{1}{2}\sum_{t=1}^{T}\left(\Delta y_{j,t} - \Delta\mu_t - \sqrt{V_j}\Delta\psi_{j,t}\right)^2\right)\right] \\
&\times \prod_{j=1}^{J} V_j^{-a_{V_j}-1}\exp\left[-\frac{1}{V_j}b_{V_j}\right]
\end{aligned}$$

where we define $\Delta x_{j,t} = x_{j,t} - x_{j,t-1}$ for $t = 2,3,\ldots,T$ and $\Delta x_{j,1} = x_{j,1}$ for any variable $x_{j,t}$ except in the case of $x_{j,t} = y_{j,t}$ where we define $\Delta y_{j,1} = y_{j,1} - \psi_{j,0}$. This allows us to write the model as

$$y_{j,t} = y_{j,t-1} + \sqrt{V_j}\Delta\psi_{j,t} + u_t + w_{j,t}, \tag{7.6}$$

where we define $y_{j,0} = (\sqrt{V_j} - 1)\psi_{j,0}$ and where $(w_{1:J,1:T}, \psi_{1:J,1:T}, u_{1:T})$ are mutually independent with $\psi_{j,t} \sim N(0,1)$, $w_{j,t} \sim N(0,W_j)$, and $u_t \sim N(0,U)$ for $j = 1,2,\ldots,J$ and $t = 1,2,\ldots,T$. While the model is no longer a state space model under this parameterization, it can be viewed as a state space model for the $\Delta y_{j,t}$'s with latent states $\Delta\psi_{j,t}$'s and $u_t = \Delta\mu_t$ so long as care is taken in defining the initial values of the data and states. We did not explore this parameterization, mainly because the scaled disturbances and scaled errors are natural opposites in the sense tending to yield efficient DA algorithms in opposite ends of the parameter space [23], and as such are desirable candidates for interweaving.

Similar to the scaled disturbances case, we have

$$p(V_{1:T}, W_{1:J}, U|\psi_{0:T}, y_{1:T}) = p(U|\psi_{0:T}, y_{1:T}) \prod_{j=1}^{J} p(V_j, W_j|\psi_{0:T}, y_{1:T}).$$

Once again $p(U|\psi_{0:T}, y_{1:T}) = p(U|\theta_{0:T}, y_{1:T})$, which is the same inverse-gamma draw. In fact, the parameters $\tilde{a}_U$ and $\tilde{b}_U$ do not change from the γ step to the ψ step, so the second draw of U is redundant and can be removed from the algorithm. The conditional density $p(V_j, W_j|\psi_{0:T}, y_{1:T})$ is once again complicated and has the same form as $p(W_j, V_j|\gamma_{0:T}, y_{1:T})$, i.e. it switches the positions of V_j and W_j. So again we draw V_j and W_j in separate Gibbs steps, and $W_j|V_j, \psi_{0:T}, y_{1:T}$ has the same inverse-gamma density as $W_j|\theta_{0:T}, y_{1:T}$. The density of $V_j|W_j, \psi_{0:T}, y_{1:T}$ has the form

$$p(V_j|W_j\psi_{0:T}, y_{1:T}) \propto V_j^{-a_{V_j}-1} \exp\left[-b_{V_j}\frac{1}{V_j} + c_{V_j}\sqrt{V_j} - d_{V_j}V_j\right],$$

where

$$c_{V_j} = \frac{\sum_{t=1}^{T} \Delta\psi_{j,t}(\Delta y_{j,t} - \Delta\mu_t)}{W_j} \in \Re\,, \qquad d_{V_j} = \frac{\sum_{t=1}^{T} (\Delta\psi_{j,t})^2}{2W_j} > 0.$$

This density has the same form as $p(W_j|V_j, \gamma_{0:T}, y_{1:T})$ so the same rejection sampling strategy can be used to sample from it.

Finally, we can write the GIS algorithm as follows:

1. Draw $\theta_{0:T} \sim N(\Omega^{-1}\omega, \Omega^{-1})$ using the MCFA.
2. Draw $U \sim IG(\tilde{a}_U, \tilde{b}_U)$.
3. For $j = 1,2,\ldots,J$:
   a. Draw $V_j \sim IG(\tilde{a}_{V_j}, \tilde{b}_{V_j})$
   b. Transform $\theta_{j,0:T} \to \gamma_{j,0:T}$ and draw $W_j \sim p(W_j|V_j, \gamma_{0:T}, y_{1:T})$.
   c. Transform $\gamma_{j,0:T} \to \psi_{j,0:T}$ and draw $V_j \sim p(V_j|W_j, \psi_{0:T}, y_{1:T})$.
   d. Draw $W_j \sim IG(\tilde{a}_{W_j}, \tilde{b}_{W_j})$.

Since $(U, V_1, \ldots, V_J, W_1, \ldots, W_J)$ are conditionally independent in the posterior no matter which of the DAs we use, Step 3 can be parallelized and step 2 can come before or after step 3, though we did not experiment with these possibilities. Steps

3.b and 3.c can both be accomplished using the rejection sampling method described in Appendix E of [23], briefly described above. Note that the transformation from $\gamma_{j,0:T} \rightarrow \psi_{j,0:T}$ is defined as $\psi_{j,t} = (y_{j,t} - \mu_t - \sqrt{W}_j \sum_{s=1}^{t} \gamma_{j,s} - \gamma_{j,0})/\sqrt{V_j}$ for $j = 1, 2, \ldots, J$ and $t = 1, 2, \ldots, T$.

In (7.5) and (7.6), it is apparent that using the scaled disturbances or the scaled errors, the model no longer has the Markov property. This is undesirable for computational reasons—it causes the double summations in the definitions of c_{W_i} and d_{W_i} and increases the computational cost associated with drawing the latent states—but the cost is worthwhile for convergence and mixing because the parameterizations are natural opposites in a particular sense. According to both theorem 1 and theorem 2 of [27], the convergence rate of an interweaving algorithm is faster when the convergence rate of the fastest underlying DA algorithm is faster, so in their words it is desirable to seek a "beauty and the beast" pair of DAs where when one DA algorithm is bad the other is good and vice-versa. Simpson et al. [23] showed in the local level model that the scaled disturbances and scaled errors yield DA algorithms which are efficient in opposite ends of the parameter space so that they exhibit precisely this "beauty and the beast" behavior.

It is also possible to transform the μ_t's in an interweaving approach. The problem becomes which two parameterizations to use. The scaled disturbances and the scaled errors make a natural pair because they work well in opposite ends of the parameter space which, in turn, seems to be driven by one being a data level reparameterization and the other a latent state level reparameterization. The scaled version of the μ_t's would still be a latent state level parameterization, and there is no clear data level reparameterization which corresponds to them. This is a consequence of the model having a higher dimensional latent state than data, though one method to overcome this issue that [23] mentions is via additional augmentation—that is, define missing data on the data level so that the full data, consisting of the observed and missing data, has the same dimension as the latent state. We sidestep this issue by leaving the μ_t's untransformed through the algorithm, though there are potential gains to be made by experimenting with reparameterizing this component of the DA.

7.5 Results

We fit the model in R using both MCMC algorithms, running five chains for each algorithm at diverse starting points for 20,000 iterations per chain. For both algorithms, convergence appeared to be attained for all parameters in all chains in the first 5,000 iterations according to both trace plots and the Gelman–Rubin diagnostic [2], so we throw away those initial draws as burn in. The GIS algorithm appeared to converge slightly slower according to the Gelman–Rubin diagnostic for some of the parameters, though this difference was not apparent in trace plots.

There were, however, significant differences in mixing between the two algorithms. Table 7.1 contains the effective sample size, n_{eff} [11], for each parameter as well as the time in seconds to achieve an effective sample size of 1,000 for each

Table 7.1 Effective sample size (n_{eff}) and time in seconds per 1,000 effective draws (Time) for each MCMC algorithm computed after burn-in for all chains. Actual sample size is 60,000 for each algorithm

	V_1	V_2	V_3	V_4	V_5	V_6	U	W_1	W_2	W_3	W_4	W_5	W_6
DA n_{eff}	24,633	20,656	20,558	18,883	21,003	24,897	14,583	15,072	18,713	15,137	10,609	13,228	29,458
GIS n_{eff}	44,894	43,659	35,400	43,843	23,364	40,913	19,571	23,706	23,560	22,768	15,051	17,753	29,729
DA time	3.08	3.68	3.70	4.02	3.62	3.05	5.21	5.04	4.06	5.02	7.16	5.74	2.58
GIS time	4.85	4.98	6.15	4.96	9.31	5.32	11.12	9.18	9.24	9.56	14.46	12.26	7.32

parameter, computed for both MCMC algorithms using all 60,000 post burn-in iterations. The GIS algorithm has higher n_{eff} for all parameters. For some parameters, e.g. V_5 and W_6, this difference is rather small. For others, such as V_1 and V_2, the GIS algorithm has an n_{eff} roughly twice as large as the DA algorithm. In time per 1,000 effective draws, however, the GIS algorithm under-performs across the board. When evaluating these times, note that the algorithms were implemented in R where the code was interpreted, not compiled. Absolute times may differ dramatically from the times listed in Table 7.1 under different programming languages or based on whether the code was interpreted or compiled, though relative times should be roughly comparable at least for interpreted code from other languages. The steps to draw from $p(W_j|V_j, \gamma_{0:T}, y_{1:T})$ and $p(V_j|W_j, \psi_{0:T}, y_{1:T})$ are the main culprits, as they are often very expensive. As the number of periods in the experiment increases, [23] found that in the local level model the GIS algorithm looks stronger relative to the DA algorithm since GIS is able to use adaptive rejection sampling more often and the relative advantage of the improved mixing becomes more important, and we expect this to hold in our model. Similarly, a judicious choice of priors which allows for easier full conditionals in the offending steps should result in a faster computational times for GIS relative to the DA algorithm.

Table 7.2 contains the parameter estimates for the model. The treatment level variance appears to be smaller than both the replication and observation level variances, suggesting that changes in logit efficiency over time are driven less by treatment level dynamics and more by random noise and replication level dynamics. Figure 7.1 also contains plots of each replication's observed logit efficiency trajectory, each replication's posterior median latent logit efficiency trajectory, and the treatment wide posterior median latent efficiency trajectory. The replication level latent logit efficiency follows the observed logit efficiency very closely in each case—it is essentially a smoothed version of the observed logit efficiency. The treatment latent logit efficiency follows the observed logit efficiencies of replications 2, 4, 5, and 6 fairly closely, but replication 3 consistently underperforms the treatment average while replication 1 consistently over performs, at least in the latter half of periods.

Table 7.2 Parameter estimates, including the posterior mean, posterior median, and a 95 % credible interval for each parameter

	Mean	50 %	2.5 %	97.5 %		Mean	50 %	2.5 %	97.5 %
V_1	0.144	0.136	0.070	0.263	W_1	0.101	0.092	0.042	0.216
V_2	0.086	0.080	0.040	0.163	W_2	0.083	0.075	0.035	0.171
V_3	0.116	0.106	0.045	0.248	W_3	0.078	0.072	0.035	0.158
V_4	0.102	0.095	0.046	0.196	W_4	0.104	0.095	0.043	0.216
V_5	0.208	0.196	0.075	0.415	W_5	0.110	0.096	0.038	0.258
V_6	0.162	0.153	0.077	0.296	W_6	0.085	0.076	0.034	0.188
					U	0.044	0.041	0.023	0.079

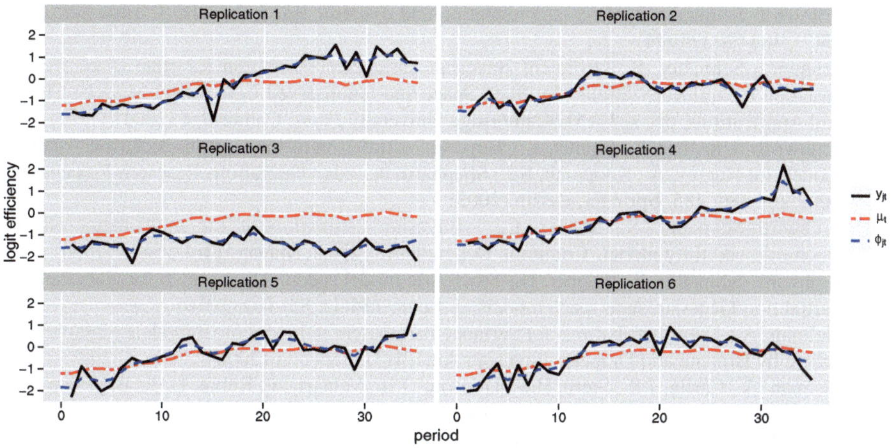

Fig. 7.1 Plots by replication of the observed logit efficiency ($y_{j,t}$), posterior median latent replication logit efficiency ($\phi_{j,t}$), and posterior median latent treatment logit efficiency (μ_t)

7.6 Conclusion

Simpson et al. [23] explored the interweaving algorithms of [27] for DLMs, but only implemented them in the univariate local level model. We use their approach in a model that can be represented as independent local level models conditional on a univariate sequence of latent states, or as a slightly more complicated DLM with J-dimensional data and $J+1$-dimensional state. This poses some problems with directly applying the methods in [23], but we show that they are easily overcome. The resulting sampler has similar convergence and improved mixing properties compared to the standard data augmentation algorithm with this particular dataset. In terms of end user time required to adequately characterize the posterior, the DA algorithm is a bit faster for this particular problem despite worse mixing, but this is largely due to an inefficient rejection sampling step in the interweaving algorithm that likely can be improved [23]. This step also tends to become relatively more efficient in problems with more data as well as less important relative to improved mixing so that the interweaving algorithm will eventually, with enough data, outperform the DA algorithm [23].

References

[1] Bos, C.S., Shephard, N.: Inference for adaptive time series models: Stochastic volatility and conditionally Gaussian state space form. Econ. Rev. **25**(2–3), 219–244 (2006)

[2] Brooks, S.P., Gelman, A.: General methods for monitoring convergence of iterative simulations. J. Comput. Graph. Stat. **7**(4), 434–455 (1998)

[3] Carter, C.K., Kohn, R.: On Gibbs sampling for state space models. Biometrika **81**(3), 541–553 (1994)

[4] Crockett, S., Smith, V.L., Wilson, B.J. Exchange and specialisation as a discovery process. Econ. J. **119**(539), 1162–1188 (2009)

[5] Frühwirth-Schnatter, S.: Data augmentation and dynamic linear models. J. Time Ser. Anal. **15**(2), 183–202 (1994)
[6] Frühwirth-Schnatter, S.: Efficient Bayesian parameter estimation for state space models based on reparameterizations. In: State Space and Unobserved Component Models: Theory and Applications, pp. 123–151. Cambridge University Press, Cambridge (2004)
[7] Frühwirth-Schnatter, S., Sögner, L.: Bayesian estimation of the Heston stochastic volatility model. In: Harvey, A., Koopman, S.J., Shephard, N. (eds.) Operations Research Proceedings 2002, pp. 480–485. Springer, Berlin (2003)
[8] Frühwirth-Schnatter, S., Sögner, L.: Bayesian estimation of the multi-factor Heston stochastic volatility model. Commun. Dependability Qual. Manag. **11**(4), 5–25 (2008)
[9] Frühwirth-Schnatter, S., Wagner, H.: Stochastic model specification search for Gaussian and partial non-Gaussian state space models. J. Econ. **154**(1), 85–100 (2010)
[10] Gelman, A.: Prior distributions for variance parameters in hierarchical models (comment on article by Browne and Draper). Bayesian Anal. **1**(3), 515–534 (2006)
[11] Gelman, A., Carlin, J.B., Stern, H.S., Dunson, D.B., Vehtari, A., Rubin, D.B.: Bayesian Data Analysis, 3rd edn. CRC press, New York (2013)
[12] Gilks, W.R., Wild, P.: Adaptive rejection sampling for Gibbs sampling. Appl. Stat. **41**(2), 337–348 (1992)
[13] Huang, A., Wand, M.P.: Simple marginally noninformative prior distributions for covariance matrices. Bayesian Anal. **8**(2), 439–452 (2013)
[14] Kaldor, N.: Welfare propositions of economics and interpersonal comparisons of utility. Econ. J. **49**, 549–552 (1939)
[15] Kastner, G., Frühwirth-Schnatter, S.: Ancillarity-sufficiency interweaving strategy (ASIS) for boosting MCMC estimation of stochastic volatility models. Comput. Stat. Data Anal. **76**, 408–423 (2014)
[16] Kimbrough, E.O., Smith, V.L., Wilson, B.J.: Exchange, theft, and the social formation of property. J. Econ. Behav. Organ. **74**(3), 206–229 (2010)
[17] Mas-Colell, A., Whinston, M.D., Green, J.R., et al.: Microeconomic Theory, vol. 1. Oxford university press, New York (1995)
[18] McCausland, W.J., Miller, S., Pelletier, D.: Simulation smoothing for state–space models: a computational efficiency analysis. Comput. Stat. Data Anal. **55**(1), 199–212 (2011)
[19] Pitt, M.K., Shephard, N.: Analytic convergence rates and parameterization issues for the Gibbs sampler applied to state space models. J. Time Ser. Anal. **20**(1), 63–85 (1999)
[20] Roberts, G.O., Papaspiliopoulos, O., Dellaportas, P.: Bayesian inference for non-Gaussian Ornstein–Uhlenbeck stochastic volatility processes. J. R. Stat. Soc. Ser. B Stat. Methodol. **66**(2), 369–393 (2004)
[21] Rue, H.: Fast sampling of Gaussian markov random fields. J. R. Stat. Soc. Ser. B Stat. Methodol. **63**(2), 325–338 (2001)
[22] Shephard, N.: Statistical Aspects of ARCH and Stochastic Volatility. Springer, London (1996)
[23] Simpson, M., Niemi, J., Roy, V.: Interweaving Markov chain Monte Carlo strategies for efficient estimation of dynamic linear models. Working Paper (2014)
[24] Strickland, C.M., Martin, G.M., Forbes, C.S.: Parameterisation and efficient MCMC estimation of non-Gaussian state space models. Comput. Stat. Data Anal. **52**(6), 2911–2930 (2008)
[25] Tanner, M.A., Hung Wong, W.: The calculation of posterior distributions by data augmentation. J. Am. Stat. Assoc. **82**(398), 528–540 (1987)
[26] Van Dyk, D., Meng, X.L.: The art of data augmentation. J. Comput. Graph. Stat. **10**(1), 1–50 (2001)
[27] Yu, Y., Meng, X.L.: To center or not to center: That is not the question - an ancillarity–sufficiency interweaving strategy (ASIS) for boosting MCMC efficiency. J. Comput. Graph. Stat. **20**(3), 531–570 (2011)

Chapter 8
On Bayesian Based Adaptive Confidence Sets for Linear Functionals

Botond Szabó

Abstract We consider the problem of constructing Bayesian based confidence sets for linear functionals in the inverse Gaussian white noise model. We work with a scale of Gaussian priors indexed by a regularity hyper-parameter and apply the data-driven (slightly modified) marginal likelihood empirical Bayes method for the choice of this hyper-parameter. We show by theory and simulations that the credible sets constructed by this method have sub-optimal behaviour in general. However, by assuming "self-similarity" the credible sets have rate-adaptive size and optimal coverage. As an application of these results we construct L_∞-credible bands for the true functional parameter with adaptive size and optimal coverage under self-similarity constraint.

Key words: Credible bands, Empirical Bayes, Minimax, Coverage, Adaptation, Linear functionals

8.1 Introduction

Uncertainty quantification is highly important in statistical inference. Point estimators without confidence statements contain only a limited amount of information. Bayesian techniques provide a natural and computationally advantageous way to quantifying uncertainty by producing credible sets, i.e. sets with prescribed (typically 95 %) posterior probability. In this paper, we investigate the validity of such sets from a frequentist perspective. We are interested in the question whether these sets can indeed be used as confidence sets or by doing so one gives a misleading uncertainty quantification, see, for instance, [12]. In our work, we focus on credible sets for linear functionals in nonparametric models and their application to the construction of L_∞-credible bands for the functional parameter.

B. Szabó (✉)
Stochastics Department, Budapest University of Technology, Budapest, Hungary
e-mail: bszabo@math.bme.hu

S. Frühwirth-Schnatter et al. (eds.), *Bayesian Statistics from Methods to Models and Applications*, Springer Proceedings in Mathematics & Statistics 126,
DOI 10.1007/978-3-319-16238-6_8

In infinite-dimensional models, nonparametric priors usually have a tuning- or hyper-parameter controlling the fine details of the prior distribution. In most of the cases, the choice of the hyper-parameter in the prior distribution is very influential, incorrect choices can result in sub-optimal behaviour of the posterior. Therefore, data-driven, adaptive Bayesian techniques are applied in practice to determine the value of the hyper-parameter, overcoming overly strong prior assumptions. The two (perhaps) most well-known Bayesian methods used to achieve adaptive results are the hierarchical Bayes and the empirical Bayes techniques. In our work, we focus mainly on the empirical Bayes method, but conjecture results about the hierarchical Bayes approach as well.

The frequentist properties of adaptive Bayesian credible sets were considered only in a limited number of recent papers; see [1, 18, 20, 22, 23]. The authors of these papers have shown that under a relatively mild and natural assumption on the functional parameter, i.e. the self-similarity condition, the credible sets have good frequentist coverage and optimal size in a minimax sense. However, for non-self-similar functions the credible sets provide overconfident, misleading confidence statements. In these papers mainly the L_2-norm was considered, which is the natural extension of the finite-dimensional Euclidean-norm, but for visualization purposes it is perhaps not the most appropriate choice. In practice usually the posterior credible bands are plotted, which correspond to the L_∞-norm. The frequentist properties of L_∞-credible bands were investigated in a non-adaptive setting in [6, 24]. Adaptive L_∞-credible bands were considered up to now only in the recent work [18]. In our work, we also focus on the construction of L_∞-credible bands taking a substantially different and independently developed approach from the one by [18].

In our analysis, we consider a sub-class of (possibly non-continuous) linear functionals satisfying the self-similarity condition and construct credible sets over it using empirical Bayes method on a scale of Gaussian priors with varying regularity; see also [22] for the application of this family of priors to derive L_2-credible sets for the functional parameter. We show that by slightly modifying the empirical Bayes procedure we can construct credible sets with rate-adaptive size and good coverage property for the linear functionals of the self-similar functional parameters. However, there exist certain oddly behaving functional parameters (not satisfying the self-similarity assumptions), where the empirical Bayes method provides haphazard and misleading credible sets for the linear functionals of the functional parameter. This result is in itself of independent interest, since until now the frequentist properties of credible sets in semi-parametric problems were mostly investigated in non- adaptive settings, see, for instance, [2, 5, 15, 19] and references therein.

However, perhaps the main contribution of the present paper is the application of the derived results about linear functionals to the analysis of L_∞-credible bands. We show that point evaluations of the functional parameters satisfy the self-similarity assumption on linear functionals. Therefore the above described results about credible sets for linear functionals apply also to pointwise credible sets for the functional parameter. Then, putting together these pointwise credible sets we arrive to an L_∞-credible band, which therefore has again good frequentist properties for self-similar functional parameters. This technique is essentially different than the one applied in [18] for the construction of L_∞-credible bands, where wavelet basis with spike and slab priors were considered and a weak Bernstein–von Mises theorem was proved.

The remainder of the paper is organized as follows. In Sect. 8.2 we introduce the inverse Gaussian white noise model, where we have carried out our analysis. In Sect. 8.2.1 we introduce the linear functionals we are interested in (satisfying the self-similarity constraint) and show that the point evaluations of the functional parameter satisfy this property. The construction of the empirical Bayes credible sets is given in Sect. 8.2.2. The main results of the paper are formulated in Sects. 8.2.3 and 8.2.4. In Sect. 8.3, we provide a short numerical analysis demonstrating both the positive and negative findings of the paper. The proofs of the main theorems are deferred to Sect. 8.4.

8.2 Main Result

Consider the inverse Gaussian white noise model

$$X_t = \int_0^t \mathbb{K}\theta_0(s)ds + \frac{1}{\sqrt{n}}B_t, \qquad t \in [0,1],$$

where B_t is the Brownian motion, $1/n$ is the noise level, X_t the observed signal, $\theta_0(\cdot) \in L^2[0,1]$ the unknown function of interest and $\mathbb{K} : L^2[0,1] \mapsto L^2[0,1]$ a given compact, linear, self-adjoint transformation (but we also allow $\mathbb{K} = \mathbb{I}$). From the self-adjoint property of $\mathbb{K}$ follows that its eigenvectors $\varphi_i(\cdot) : [0,1] \mapsto \mathbb{R}$ form an orthogonal basis and the compactness ensures that the corresponding eigenvalues κ_i are tending to zero. Hence, using series expansion with respect to φ_i we get the equivalent Gaussian sequence model

$$X_i = \kappa_i\theta_{0,i} + \frac{1}{\sqrt{n}}Z_i, \quad \text{for all } i = 1,2,\dots \tag{8.1}$$

where $X_i = \langle X_\cdot, \varphi_i(\cdot)\rangle$ and $\theta_{0,i} = \langle \theta_0(\cdot), \varphi_i(\cdot)\rangle$ are the series decomposition coefficients of the observation and the true function, respectively, and the random variables $Z_i = \langle B_\cdot, \varphi_i(\cdot)\rangle$ are independent and standard normally distributed. We limit ourselves to mildly ill-posed inverse problems, where

$$C^{-2}i^{-2p} \leq \kappa_i^2 \leq C^2 i^{-2p}, \tag{8.2}$$

with some fixed non-negative constant p and positive C, see [7] for the terminology.

Suppose furthermore that the unknown infinite-dimensional parameter $\theta_0 = (\theta_{0,1}, \theta_{0,2}, ..)$ belongs to a hyper-rectangle

$$\Theta^{\beta}(M) = \{\theta : \theta_i^2 i^{1+2\beta} \leq M \quad \text{for all } i = 1,2,\dots\}, \tag{8.3}$$

where β is the regularity parameter and M is the squared radius of the hyper-rectangle. The minimax estimation rate of the full parameter θ_0 is a multiple of $n^{-\beta/(1+2\beta+2p)}$, see [10].

8.2.1 Linear Functionals

In this paper, we focus on the construction of confidence sets for the (possibly unbounded) linear functionals

$$L\theta = \sum l_i \theta_i, \tag{8.4}$$

where $l = (l_1, l_2, \dots)$ is in a self-similar hyper-rectangle $L_s^q(R)$, for some q, R, j_0, K

$$L_s^q(R) = \{l \in \ell_2 : (1/R^2) j^{-1-2q} \leq \sum_{i=j}^{j+K-1} l_i^2 \leq R^2 j^{-1-2q}, \text{for all } j > j_0\}, \tag{8.5}$$

where the parameters j_0 and K are omitted from the notation. We note that, for instance, the linear functionals in the form $l_i \asymp i^{-q}$ belong to this hyper-rectangle.

We are particularly interested in the class of non-continuous linear functionals, the point evaluations of the functional parameter θ. For a specific choice of the operator $\mathbb{K}$, all point evaluations on $t \in [0,1]$ belong to $L_s^{-1/2}(R)$; see the next paragraph. Therefore, confidence sets for self-similar linear functionals $L \in L_s^{-1/2}(R)$ of the series decomposition coefficients θ_0 also provide us with pointwise confidence sets of the function $\theta_0(\cdot)$. Gluing together the (uniform) pointwise confidence sets one arrives at L_∞-confidence bands.

In this paragraph we show that point evaluations of the function $\theta_0(\cdot)$ belong to the self-similar class of linear functionals for appropriate choice of the basis. Assume that the eigen-basis of the operator $\mathbb{K}$ is the sine-cosine basis $\varphi_i(\cdot)$. The function $\theta_0(\cdot)$ can be given with the help of the trigonometric decomposition

$$\theta_0(t) = \sum_i \theta_{0,i} \varphi_i(t). \tag{8.6}$$

Since $\varphi_{2i+1}^2(t) + \varphi_{2i}^2(t) = \sin^2(i2\pi t) + \cos^2(i2\pi t) = 1$ we obtain that $l_i = \varphi_i(t)$ is in $L_s^{-1/2}(2)$ with parameters $j_0 = 1$ and $K = 3$ (since every three consecutive integers contain a pair of $(2i-1, 2i)$ for some $i \in \mathbb{N}$).

8.2.2 Bayesian Approach

In the Bayesian framework of making inference about the unknown sequence θ_0, we endow it with a prior distribution. In our analysis, we work with the infinite-dimensional Gaussian distribution

$$\Pi_\alpha = \bigotimes_{i=1}^{\infty} N(0, i^{-1-2\alpha}), \tag{8.7}$$

where the parameter $\alpha > 0$ denotes the regularity level of the prior distribution. One can easily compute the corresponding posterior distribution

$$\Pi_\alpha(\cdot|X) = \bigotimes_{i=1}^{\infty} N\Big(\frac{n\kappa_i^{-1}}{i^{1+2\alpha}\kappa_i^{-2}+n}X_i, \frac{\kappa_i^{-2}}{i^{1+2\alpha}\kappa_i^{-2}+n}\Big). \tag{8.8}$$

Furthermore by combining and slightly extending the results of [4] and [15] one can see that the choice $\alpha = \beta$ leads to the posterior contraction rate $n^{-\frac{\beta}{1+2\beta+2p}}$ for $\theta_0 \in \Theta^\beta(M)$, while other choices of the parameter α provide sub-optimal contraction rates.

In this paper, however, we are interested in the posterior distribution of the linear functional $L\theta$. From Proposition 3.2 of [15] follows that the posterior distribution of the linear functionals $L\theta$ (assuming measurability with respect to the prior Π_α) takes the form

$$\Pi_\alpha^L(\cdot|X) = N\Big(\sum_i \frac{nl_i\kappa_i^{-1}}{i^{-1-2\alpha}\kappa_i^{-2}+n}X_i, \sum_i \frac{l_i^2\kappa_i^{-2}}{i^{1+2\alpha}\kappa_i^{-2}+n}\Big). \tag{8.9}$$

Furthermore it was also shown in Section 5 of [15] that the optimal choice of the hyper-parameter α is not β, but rather $\beta - 1/2$. The resulting optimal rate is of the order $n^{-(\beta+q)/(2\beta+2p)} \vee n^{-1/2}$; see [9, 11]. Note that in case $q \geq p$ the smoothness of the linear functional compensates for the degree of ill-posedness and we get a regular problem with contraction rate $n^{-1/2}$. However, in our work we focus on the more interesting case $q < p$ from the point of view of constructing credible bands.

Since the regularity parameter β of the infinite sequence θ_0 is usually not available, one has to use data-driven methods to choose α, which from now on we will refer to as the hyper-parameter of the prior. Following [14] and [22] we select a value for α with the marginal likelihood empirical Bayes method, i.e. we select the maximizer of

$$\hat{\alpha}_n = \arg\max_{\alpha \in [0,A]} \ell_n(\alpha), \tag{8.10}$$

where A is some arbitrary large, fixed constant, and ℓ_n denotes the corresponding log-likelihood for α (relative to an infinite product of $N(0, 1/n)$-distributions):

$$\ell_n(\alpha) = -\frac{1}{2}\sum_{i=1}^{\infty}\Big(\log\Big(1+\frac{n}{i^{1+2\alpha}\kappa_i^{-2}}\Big) - \frac{n^2}{i^{1+2\alpha}\kappa_i^{-2}+n}X_i^2\Big). \tag{8.11}$$

Then the *empirical Bayes posterior* for the functional parameter θ is defined as $\Pi_{\hat{\alpha}_n}(\cdot|X)$ which is obtained by substituting $\hat{\alpha}_n$ for α in the posterior distribution (8.8), i.e.

$$\Pi_{\hat{\alpha}_n}(B|X) = \Pi_\alpha(B|X)\Big|_{\alpha=\hat{\alpha}_n}$$

for measurable subsets $B \subset \ell^2$. Slightly adapting the proof of Theorem 2.3 in [14] we can derive that the posterior distribution of the functional parameter θ achieves the corresponding minimax contraction rate up to a logarithmic factor.

As conjectured in page 2367 of [15] and Section 2.3 of [14] this suggests that the present procedure is sub-optimal for the linear functional $L\theta_0$, since adaptation for the full parameter θ_0 and its linear functionals $L\theta_0$ is not possible simultaneously in this setting. In view of the findings in the non-adaptive case [15] we might expect, however, that we can slightly alter the procedures to deal with linear functionals. For instance, it is natural to expect that the empirical Bayes posterior for linear functionals $L\theta$, given in (8.4),

$$\Pi^L_{\hat{\alpha}_n-1/2}(\cdot\,|X) = \Pi^L_\alpha(\cdot\,|X)\Big|_{\alpha=\hat{\alpha}_n-1/2}, \tag{8.12}$$

yields optimal rates. In the present paper we work with this data-driven choice of the hyper-parameter and investigate the frequentist properties of Bayesian credible sets constructed from the posterior (8.12).

For fixed hyper-parameter α the posterior in (8.9) is a one-dimensional Gaussian distribution hence a natural choice of the credible set is the interval

$$\hat{C}_{n,\alpha} = [\widehat{L\theta}_{n,\alpha} - \zeta_{\gamma/2}s_n(\alpha), \widehat{L\theta}_{n,\alpha} + \zeta_{\gamma/2}s_n(\alpha)], \tag{8.13}$$

where $\widehat{L\theta}_{n,\alpha}$ is the posterior mean, $s_n^2(\alpha)$ the posterior variance given in (8.9), and ζ_γ is the $(1-\gamma)$-quantile of the standard normal distribution. We note that the mean $\widehat{L\theta}_{n,\alpha}$ of the posterior distribution of the linear functional is exactly the linear functional L of the posterior mean of the full parameter $\hat{\theta}_{n,\alpha}$ given in (8.8). One can easily see that the preceding interval accumulates a $(1-\gamma)$ fraction of the posterior mass. Then the empirical Bayes credible sets are obtained by replacing α with the data-driven choice $\hat{\alpha}_n - 1/2$ in (8.13). We introduce some additional flexibility, by allowing the blow up of the preceding interval with a constant factor $D > 0$, i.e.

$$\hat{C}_n^L(D) = [\widehat{L\theta}_{n,\hat{\alpha}_n-1/2} - D\zeta_{\gamma/2}s_n(\hat{\alpha}_n-1/2), \widehat{L\theta}_{n,\hat{\alpha}_n-1/2} + D\zeta_{\gamma/2}s_n(\hat{\alpha}_n-1/2)]. \tag{8.14}$$

8.2.3 Negative Results

However, matters seem to be more delicate than one would expect from the non-adaptive case. For certain oddly behaving functions the posterior distribution (8.12) achieves only sub-optimal contraction rates. Furthermore, the credible sets (8.14) have coverage tending to zero.

Theorem 1. *Let n_j be positive integers such that $n_1 \geq 2$ and $n_j \geq n_{j-1}^4$ for every j, and let $K > 0$. Let $\theta_0 = (\theta_{0,1}, \theta_{0,2}, \dots)$ be such that*

$$\theta_{0,i} = \begin{cases} K n_j^{-\frac{1/2+\beta}{1+2\beta+2p}}, & \text{if } n_j^{\frac{1}{1+2\beta+2p}} \leq i < 2n_j^{\frac{1}{1+2\beta+2p}}, \qquad j = 1, 2, \dots, \\ 0, & \text{else}, \end{cases} \tag{8.15}$$

for some positive constants β, p. Then the constant $K > 0$ can be chosen such that the coverage of the credible set tends to zero for every $q \in \mathbb{R}$, $D, R > 0$ and $L \in L_s^q(R)$:

$$\mathrm{P}_{\theta_0}(\theta_0 \in \hat{C}_{n_j}(D)) \to 0.$$

Furthermore, the posterior distribution attains sub-optimal contraction rate

$$\Pi^L_{\hat{\alpha}_{n_j}-1/2}(L\theta : |L\theta_0 - L\theta| \geq m n_j^{-(\beta+q)/(1+2\beta+2p)} | X) \stackrel{\mathrm{P}_{\theta_0}}{\to} 1, \tag{8.16}$$

as $j \to \infty$ for a positive, small enough constant m and linear functional L satisfying (8.5).

The proof of Theorem 1 is given in Sect. 8.4.1. The sub-optimal contraction rate of the posterior distribution and the bad coverage property of the credible sets are due to the mismatch of the underlying loss functions. In the empirical Bayes method, the hyper-parameter α is chosen to maximize the marginal likelihood function. This method is related to minimizing the Kullback–Leibler divergence between the marginal Bayesian likelihood function and the true likelihood function. At the same time, the evaluation of the posterior distribution is given with respect to some linear functional L of the functional parameter θ_0. Optimal contraction rate and good coverage follow from optimal bias-variance trade-off. However, the likelihood based empirical Bayes method intends to minimize the Kullback–Leibler divergence, which is not an appropriate approach in general for balancing out the bias and variance terms. Therefore, the empirical Bayes (and we believe that also the hierarchical Bayes method) leads to sub-optimal rate and poor coverage.

8.2.4 Self-Similarity

To solve this problem, we can introduce some additional constraint on the regularity class $\Theta^{\beta}(M)$. The notion of self-similarity originates from the frequentist literature [3, 8, 13, 16, 17], and was adapted in the Bayesian literature [1, 18, 22]. We call a series in the hyper-rectangle $\theta_0 \in \Theta^{\beta}(M)$ *self-similar* if it satisfies

$$\sum_{i=N}^{\rho N} \theta_{0,i}^2 \geq \varepsilon M N^{-2\beta}, \quad \forall N \geq N_0, \tag{8.17}$$

where N_0, ε and ρ are some fixed positive constants. Furthermore we denote the class of functions satisfying the self-similar constraint by $\Theta_s^{\beta}(M)$, where we omit the parameters N_0, ε, ρ from the notation. We denote by $\Theta_s(M)$ the collection of self-similar functions with regularity in a compact interval of regularity parameters $\beta \in [\beta_{\min}, \beta_{\max}]$:

$$\Theta_s(M) = \cup_{\beta \in [\beta_{\min}, \beta_{\max}]} \Theta_s^{\beta}(M). \tag{8.18}$$

Here, we omit again the dependence on $\beta_{\min}$ and $\beta_{\max}$ in the notation and assume that $\beta_{\min} > -q$, else $L\theta_0$ would be infinite.

We show that uniformly over $L \in L_s^q(R)$ and $\theta_0 \in \Theta_s(M)$ the coverage of credible sets $\hat{C}_n^L(D)$ for the linear functionals $L\theta_0$ tends to one. Furthermore we prove that the size of the credible sets achieves the corresponding minimax contraction rate.

Theorem 2. *There exists a large enough positive constant D such that the empirical Bayes credible sets $\hat{C}_n^L(D)$ have honest asymptotic coverage one over the self-similar linear functionals $L \in L_s^q(R)$ of the functional parameter θ_0 satisfying* (8.5), *i.e.*

$$\inf_{\theta_0 \in \Theta_s(M)} P_{\theta_0}\big(L\theta_0 \in \hat{C}_n^L(D), \forall L \in L_s^q(R)\big) \to 1. \tag{8.19}$$

Furthermore, the radius of the credible sets is rate adaptive, i.e. there exists a positive constant $C_1 > 0$ such that for all $\beta \in (q, \beta_{\max}]$ we have

$$\inf_{\theta_0 \in \Theta_s^{\beta}(M)} \mathrm{P}_{\theta_0}\big(s_n(\hat{\alpha}_n - 1/2) \leq C_1 n^{-\frac{\beta+q}{2\beta+2p}}, \forall L \in L_s^q(R)\big) \to 1. \tag{8.20}$$

We defer the proof to Sect. 8.4.2. The credible band on $[0,1]$ can be constructed with the help of the linear functionals $\varphi_i(t)$ introduced in (8.6), i.e. the point evaluations of the basis $\varphi_i(\cdot)$ at $t \in [0,1]$. Following from its definition (8.14) the credible band takes the form

$$[\hat{\theta}_n(t) - D\zeta_{\gamma/2} s_n(t, \hat{\alpha}_n - 1/2), \hat{\theta}_n(t) + D\zeta_{\gamma/2} s_n(t, \hat{\alpha}_n - 1/2)], \quad t \in [0,1], \tag{8.21}$$

where $\hat{\theta}_n(t)$ is the posterior mean and $s_n^2(t, \hat{\alpha}_n - 1/2)$ is the posterior variance for $\alpha = \hat{\alpha}_n - 1/2$ given in (8.12) belonging to the linear functional $L = (l_i)_{i\geq 1} = (\varphi_i(t))_{i\geq 1}$. By combining Theorem 2 and the argument given in the last paragraph of Sect. 8.2.1 we get that the credible band (8.21) has honest coverage and rate-adaptive size.

Corollary 1. *Assume that the eigen-vectors $\varphi_i(\cdot)$ of the linear operator $\mathbb{K}$ form the sine-cosine basis. Then there exists a constant D such that the empirical Bayes credible bands, given in* (8.21)*, have honest asymptotic coverage one*

$$\inf_{\theta_0\in\Theta_s(M)} P_{\theta_0}\left(|\theta_0(t) - \hat{\theta}_{n,\hat{\alpha}_n-1/2}(t)| \leq D\zeta_{\gamma/2}s_n(t, \hat{\alpha}_n - 1/2), \forall t \in [0,1]\right) \to 1. \tag{8.22}$$

Furthermore, the size of the credible band is rate optimal in a minimax sense, i.e. there exists a $C_1 > 0$ such that for all $\beta \in (1/2, \beta_{\max}]$

$$\inf_{\theta_0\in\Theta_s^\beta(M)} \mathrm{P}_{\theta_0}\left(s_n(t, \hat{\alpha}_n - 1/2) \leq C_1 n^{-\frac{\beta-1/2}{2\beta+2p}}\right) \to 1. \tag{8.23}$$

8.3 Simulation Study

We investigate our new empirical Bayes method in an example. We consider the model (8.1) with $\mathbb{K} = \mathbb{I}$ (the identity operator) and work with the sine-cosine basis on $[0,1]$, i.e. $\varphi_1(t) = 1$, $\varphi_{2i}(t) = \sqrt{2}\cos(2\pi it)$, $\varphi_{2i+1}(t) = \sqrt{2}\sin(2\pi it)$ for $t \in [0,1]$.

First, we illustrate that for self-similar functions our method provides reasonable and trustworthy credible sets which could be used as confidence bands. We define the true function $\theta_0(t)$ with the help of its sine-cosine basis coefficients, i.e. we take $\theta_{0,i} = i^{-2}\cos(i)$:

$$\theta_0(t) = \cos(1) + \sqrt{2}\sum_{i=1}^{\infty}(2i)^{-2}\cos(2i)\cos(2\pi it)$$
$$+ \sqrt{2}\sum_{i=1}^{\infty}(2i+1)^{-2}\cos(2i+1)\sin(2\pi it).$$

For computational convenience we work only with the first 10^3 Fourier coefficients of the true function. We simulate data from the corresponding distribution with noise level $n = 100, 10^3, 10^4$ and 10^5. Figure 8.1 shows the true function in pointed black, the posterior mean in dashed red and the 95 % credible bands (without blowing it up by a constant factor D) in blue. One can see that for every noise level n the credible band has good coverage and is concentrating around the truth as n increases, confirming the results of Corollary 1.

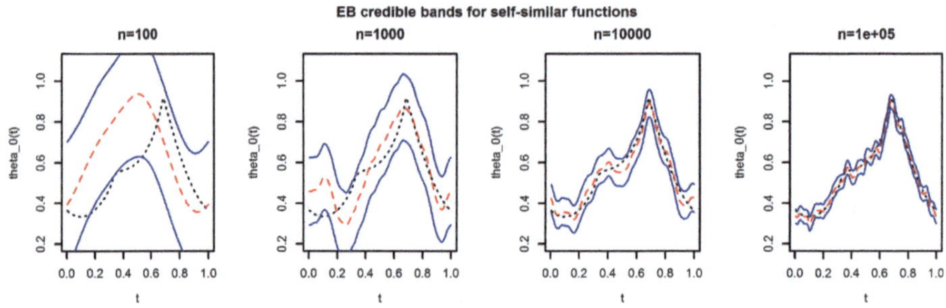

Fig. 8.1 Empirical Bayes credible bands for a self-similar function. The true function is drawn in *pointed black*, the posterior mean in *dashed red* and the credible bands in *blue*. From *left* to *right* we have $n = 100, 10^3, 10^4$ and 10^5

To illustrate the negative result derived in Theorem 1 (for the point evaluation linear functionals) we consider a non-self-similar function $\theta_0(t)$ defined by its series decomposition coefficients with respect to the sine-cosine basis. We take the coefficients to be $\theta_{0,1} = 1/10$, $\theta_{0,4} = 1/30$, $\theta_{0,20} = -1/20$, $\theta_{0,i} = i^{-3/2}$ if $2^{4^j} < i \leq 2 \cdot 2^{4^j}$ for $j \geq 2$, and 0 otherwise:

$$\theta_0(t) = 0.1 + \frac{\sqrt{2}}{30}\cos(4\pi t) - \frac{\sqrt{2}}{20}\cos(30\pi t)$$
$$+ \sum_{j=2}^{\infty}\Big(\sqrt{2}\sum_{i=2^{4^j-1}+1}^{2^{4^j}}(2i)^{-3/2}\cos(2\pi i t) + (2i+1)^{-3/2}\sin(2\pi i t)\Big).$$

For simplicity we consider again only the first 10^3 Fourier coefficients of the true function. Then we simulate data from the corresponding distribution with various noise levels $n = 200, 500, 10^3, 2 \cdot 10^3, 5 \cdot 10^3, 10^4, 10^5$ and 10^8. In Fig. 8.2 we plotted the 95 % L_∞-credible bands with blue lines, the posterior mean with dashed red line and the true function with pointed black line. One can see that for multiple noise levels we have overly confident, too narrow credible bands ($n = 500, 10^3, 2 \cdot 10^3, 10^4$), while for other values of the noise levels n we have good coverage ($n = 200, 5 \cdot 10^3, 5 \cdot 10^4, 10^8$). This periodicity between the good and the bad coverage of the credible sets continuous as n increases (but to see it we have to zoom into the picture).

8.4 Proofs and Lemma

8.4.1 Proof of Theorem 1

Following [14, 21] and [22] we introduce the notation

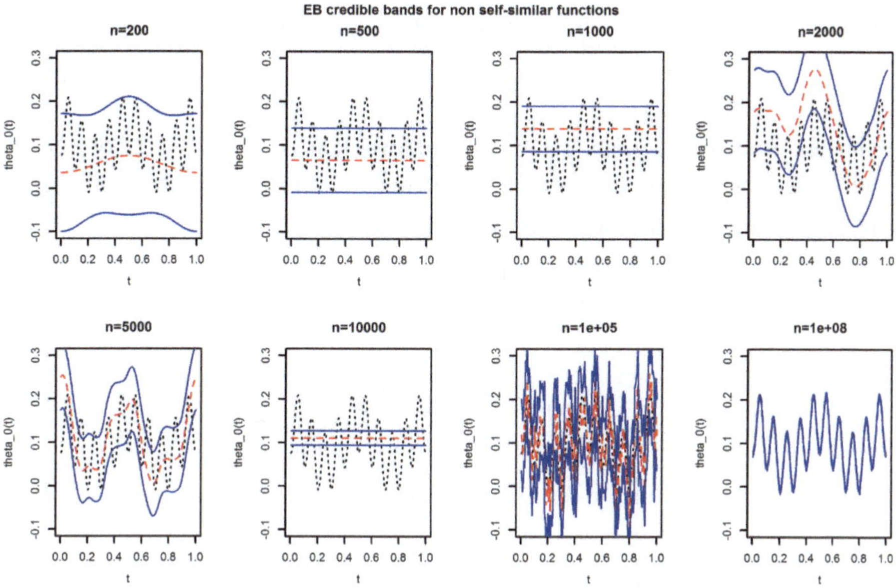

Fig. 8.2 Empirical Bayes credible bands for a non-self-similar function. The true function is drawn in *pointed black*, the posterior mean in *dashed red* and the credible bands in *blue*. From *left* to *right* and *top* to *bottom* we have $n = 200, 500, 10^3, 2\cdot 10^3, 5\cdot 10^3, 10^4, 10^5$ and 10^8

$$h_n(\alpha;\theta_0) = \frac{1+2\alpha+2p}{n^{1/(1+2\alpha+2p)}\log n}\sum_{i=1}^{\infty}\frac{n^2 i^{1+2\alpha}(\log i)\theta_{0,i}^2}{(i^{1+2\alpha+2p}+n)^2}, \qquad \alpha \geq 0,$$

and define

$$\underline{\alpha}_n(\theta_0) = \inf\{\alpha \in [0,\mathrm{A}] : h_n(\alpha;\theta_0) \geq 1/(16C^8)\},$$
$$\overline{\alpha}_n(\theta_0) = \sup\{\alpha \in [0,\mathrm{A}] : h_n(\alpha;\theta_0) \leq 8C^8\},$$

where the parameter A was introduced in (8.10). From the proof of Theorem 5.1 of [22] one can see that

$$\inf_{\theta_0\in\ell_2} \mathrm{P}(\underline{\alpha}_n \leq \hat{\alpha}_n) \to 1. \tag{8.24}$$

Furthermore, let us introduce the notations

$$B_n^L(\alpha) = |\mathrm{E}_{\theta_0}\widehat{L\theta}_\alpha - L\theta_0| \quad \text{and} \quad V_n(\alpha) = |\widehat{L\theta}_\alpha - \mathrm{E}_{\theta_0}\widehat{L\theta}_\alpha|,$$

where $\widehat{L\theta}_\alpha$ denotes the posterior mean of the linear functional $L\theta$ for a fixed hyper-parameter $\alpha > 0$. Similarly to the proof of Theorem 3.1 of [22], it follows from the triangle inequality that $\theta_0 \in \hat{C}_n^L(D)$ implies $B_n^L(\hat{\alpha}_n - 1/2) \leq V_n(\hat{\alpha}_n - 1/2) + D\zeta_{\gamma/2}s_n(\hat{\alpha}_n - 1/2)$. Therefore following from the convergence (8.24) we have

$$P_{\theta_0}(L\theta_0 \in \hat{C}_n^L(D)) \leq P_{\theta_0}\Big(\inf_{\alpha \geq \underline{\alpha}_n - 1/2} B_n(\alpha) \leq D \sup_{\alpha \geq \underline{\alpha}_n - 1/2} \big[D\zeta_{\frac{\gamma}{2}} s_n(\alpha) + V_n(\alpha)\big]\Big) + o(1). \tag{8.25}$$

From the proof of Theorem 3.1 of [22] follows that $\underline{\alpha}_{n_j} > \beta + 1/2$ for j large enough. Following from the proof of Theorem 5.3 of [15], we get that both $s_{n_j}(\alpha)$ and $V_{n_j}(\alpha)$ are bounded from above by a multiple of $n_j^{-(1/2+\beta+q)/(1+2\beta+2p)} \ll n_j^{-(\beta+q)/(2\beta+2p)}$ with probability tending to one.

Furthermore, for fixed hyper-parameter α the bias corresponding to the posterior mean (8.12) is

$$B_n^L(\alpha) = |\sum_i \frac{l_i \theta_{0,i}}{1 + n_j i^{-1-2\alpha} \kappa_i^2}|.$$

Note that following from the definition of $L_s^q(R)$ given in (8.5), we conclude that

$$\sum_{i=j}^{j+K-1} |l_i| \geq \max_{i \in \{j, j+1, \ldots, j+K-1\}} |l_i| \geq j^{-1/2-q}/(RK).$$

Then, for $\alpha \geq \underline{\alpha}_{n_j} - 1/2 \geq \beta$ the squared bias $B_n^2(\alpha)$ corresponding to the sequence (8.15) can be bounded from below by

$$\Big(\sum_{i=n_j^{\frac{1}{1+2\beta+2p}}}^{2n_j^{\frac{1}{1+2\beta+2p}}} \frac{C_0^{-1}|l_i| K n_j^{-\frac{1/2+\beta}{1+2\beta+2p}}}{1 + n_j i^{-1-2\alpha} \kappa_i^2}\Big)^2 \gtrsim n_j^{-\frac{1/2+\beta}{1+2\beta+2p}} \sum_{i=n_j^{1/(1+2\beta+2p)}/K}^{2n_j^{1/(1+2\beta+2p)}/K-1} \sum_{j=iK+1}^{(i+1)K} |l_i|$$

$$\gtrsim n_j^{-\frac{1/2+\beta}{1+2\beta+2p}} \sum_{i=n_j^{1/(1+2\beta+2p)}/K}^{2n_j^{1/(1+2\beta+2p)}/K-1} i^{-1/2-q}, \tag{8.26}$$

which is further bounded from below by $n_j^{-(2\beta+2q)/(1+2\alpha+2p)} \gg n_j^{-2(\beta+q)/(2\beta+2p)}$. Therefore, the probability on the right-hand side of the inequality (8.25) tends to zero.

Finally, we note that following from the sub-optimal order of the bias term (8.26) the posterior distribution achieves sub-optimal contraction rate around the true value $L\theta_0$.

8.4.2 *Proof of Theorem 2*

First, we note that following from the inequalities (6.9) and (6.10) of [22] we obtain for all $\beta \in [\beta_{\min}, \beta_{\max}]$ that

$$\beta - K_1/\log n \leq \inf_{\theta_0 \in \Theta^\beta(M)} \underline{\alpha}_n(\theta_0) \leq \sup_{\theta_0 \in \Theta_s^\beta(M)} \overline{\alpha}_n(\theta_0) \leq \beta + K_2/\log n, \tag{8.27}$$

for some positive constants K_1 and K_2 depending only on $\beta_{\min}, \beta_{\max}, M, C, \rho$ and ε.

Then, for convenience we introduce the notation

$$r_{n,\gamma}^2(\alpha) = D^2 \zeta_{\gamma/2}^2 s_n^2(\alpha) = D^2 \zeta_{\gamma/2}^2 \sum_{i=1}^{\infty} \frac{l_i^2 \kappa_i^{-2}}{i^{1+2\alpha}\kappa_i^{-2} + n}. \tag{8.28}$$

Using the notations of Sect. 8.4.1, the coverage of the empirical Bayes credible set, similarly to the inequality (8.25), can be bounded from below by

$$\begin{aligned} &\inf_{\theta_0 \in \Theta_s(M)} \mathrm{P}_{\theta_0}\left(L\theta_0 \in \hat{C}_n^L(D), \forall L \in L_s^q(R)\right) \qquad (8.29)\\ &\geq \inf_{\theta_0 \in \Theta_s(M)} \mathrm{P}_{\theta_0}\left(B_n^L(\hat{\alpha}_n - 1/2) + V_n(\hat{\alpha}_n - 1/2) \leq r_{n,\gamma}(\hat{\alpha}_n - 1/2), \forall L \in L_s^q(R)\right)\\ &\geq \inf_{\theta_0 \in \Theta_s(M)} \mathrm{P}_{\theta_0}\Big(\sup_{\substack{\alpha \in [\underline{\alpha}_n - 1/2, \overline{\alpha}_n - 1/2]\\ L \in L_s^q(R)}} V_n(\alpha) \leq \inf_{\substack{\alpha \in [\underline{\alpha}_n - 1/2, \overline{\alpha}_n - 1/2]\\ L \in L_s^q(R)}} \left[r_{n,\gamma}(\alpha) - B_n^L(\alpha)\right]\Big) - o(1). \end{aligned}$$

Therefore, it is sufficient to show that there exist constants C_1, C_2 and C_3 satisfying $C_3 > C_2 + C_1$ such that for all $\beta \in [\beta_{\min}, \beta_{\max}]$ and $\theta_0 \in \Theta_s^\beta(M)$

$$\sup_{\alpha \in [\underline{\alpha}_n - 1/2, \overline{\alpha}_n - 1/2]} B_n^L(\alpha) \leq C_1 n^{-\frac{\beta+q}{2\beta+2p}}, \tag{8.30}$$

$$P_{\theta_0}\Big(\sup_{\alpha \in [\underline{\alpha}_n - 1/2, \overline{\alpha}_n - 1/2]} V_n(\alpha) \leq C_2 n^{-\frac{\beta+q}{2\beta+2p}}\Big) \to 1, \tag{8.31}$$

$$\inf_{\alpha \in [\underline{\alpha}_n - 1/2, \overline{\alpha}_n - 1/2]} r_{n,\gamma}(\alpha) \geq C_3 n^{-\frac{\beta+q}{2\beta+2p}}. \tag{8.32}$$

We first deal with inequality (8.32). Applying assumptions (8.5) and (8.28) one can obtain that

$$\begin{aligned} r_{n,\gamma}^2(\alpha) &\geq \frac{D^2 \zeta_{\gamma/2}^2}{C^2} \sum_{i=j_0/K+1}^{\infty} \frac{(Ki)^{2p}}{((K+1)i)^{1+2\alpha+2p} + n} \sum_{j=iK}^{(i+1)K-1} l_j^2\\ &\geq \frac{D^2 \zeta_{\gamma/2}^2 K^{2p-2q}}{R^2 C^2 (K+1)^{1+2\alpha+2p}} \sum_{i=j_0/K+1}^{\infty} \frac{i^{-1-2q+2p}}{i^{1+2\alpha+2p} + n}, \end{aligned}$$

which following from Lemma 1 is further bounded from below by constant times $D^2 n^{-\frac{1+2\alpha+2q}{1+2\alpha+2p}}$ for $1 + 2\alpha + 2q > 0$ and infinity else. Therefore, by applying the inequality (8.27) we obtain that C_3 can be arbitrary large for a large enough choice of D^2.

Next we deal with the convergence (8.31). From the proof of Theorem 5.3 of [15] we get that $V_n(\alpha) = |t_n(\alpha)Z|$ with Z a standard normal random variable and

$$t_n^2(\alpha) = \sum_i^\infty \frac{nl_i^2\kappa_i^{-2}}{(i^{1+2\alpha}\kappa_i^{-2}+n)^2} \le R^2C^4\sum_i^\infty \frac{ni^{-1-2q+2p}}{(i^{1+2\alpha+2p}+n)^2}.$$

The right-hand side of the preceding display similarly to $s_n^2(\alpha)$ is bounded from above by constant times $n^{-\frac{1+2\alpha+2q}{1+2\alpha+2p}}$ for $1+2\alpha+2q>0$ and infinity otherwise. Then following from the inequality (8.27) one can obtain for $q<p$ that

$$\sup_{\alpha\in[\underline{\alpha}_n-1/2,\overline{\alpha}_n-1/2]} t_n(\alpha) = t_n(\underline{\alpha}_n-1/2) \lesssim n^{-\frac{\underline{\alpha}_n+q}{2\underline{\alpha}_n+2p}} \lesssim n^{-\frac{\beta+q}{2\beta+2p}},$$

providing us the convergence (8.31).

Finally, we deal with the bias term (8.30). Following from assumptions (8.3) and (8.5), we have

$$|B_n^L(\alpha)| \le \sum_{i=1}^\infty \frac{|l_i\theta_{0,i}|i^{1+2\alpha}\kappa_i^{-2}}{i^{1+2\alpha}\kappa_i^{-2}+n} \le C^2RK\sum_{i=1}^\infty \frac{i^{2\alpha+2p-\beta-q}}{i^{1+2\alpha+2p}+n}.$$

From the inequality (8.27) we have for $\alpha \ge \underline{\alpha}_n - 1/2$ and large enough n that the inequality $\beta+q<1+2\alpha+2p$ holds, hence the preceding inequality is further bounded from above by constant times $n^{-\frac{\beta+q}{1+2\alpha+2p}}$ by applying Lemma 1 (with $m=0$). So we can conclude that for $\alpha \ge \underline{\alpha}_n - 1/2 \ge \beta - 1/2 - K_1/\log n$

$$|B_n^L(\alpha)| \lesssim n^{-(\beta+q)/(1+2\alpha+2p)} \lesssim n^{-(\beta+q)/(2\beta+2p)}.$$

To prove adaptivity we note that following again from the inequality (8.27) we have

$$\sup_{\alpha\in[\underline{\alpha}_n-1/2,\overline{\alpha}_n-1/2]} s_n(\alpha) \lesssim n^{-(\underline{\alpha}_n+q)(2\underline{\alpha}_n+2p)} \lesssim n^{-(\beta+q)/(2\beta+2p)}.$$

8.4.3 Lemma 10.2 of [22]

Lemma 1 (Lemma 10.2 of [22]). *For any $l,m,r,s \ge 0$ with $c := lr-s-1>0$ and $n \ge e^{(2mr/c)\vee r}$,*

$$(3^r+1)^{-l}(\log n/r)^m n^{-c/r} \le \sum_{i=1}^\infty \frac{i^s(\log i)^m}{(i^r+n)^l} \le (3+2c^{-1})(\log n/r)^m n^{-c/r}.$$

suggesting changes, which improved the quality of the manuscript.

Acknowledgements The author would like to thank Harry van Zanten, Bartek Knapik and an anonymous referee for their suggestions leading to better readability and presentation of the manuscript.

References

[1] Belitser, E.: On coverage and oracle radial rate of DDM-credible sets under excessive bias restriction. ArXiv e-prints (2014)
[2] Bickel, P.J., Kleijn, B.J.K.: The semiparametric bernstein–von mises theorem. Ann. Stat. **40**(1), 206–237 (2012)
[3] Bull, A.: Honest adaptive confidence bands and self-similar functions. Electron. J. Stat. **6**, 1490–1516 (2012)
[4] Castillo, I.: Lower bounds for posterior rates with Gaussian process priors. Electron. J. Stat. **2**, 1281–1299 (2008)
[5] Castillo, I.: A semiparametric bernstein–von mises theorem for gaussian process priors. Probab. Theory Relat. Fields **152**(1–2), 53–99 (2012)
[6] Castillo, I., Nickl, R.: On the bernstein–von mises phenomenon for nonparametric bayes procedures. Ann. Stat. **42**(5), 1941–1969 (2014)
[7] Cavalier, L.: Nonparametric statistical inverse problems. Inverse Prob. **24**(3), 034004, 19 (2008)
[8] Chernozhukov, V., Chetverikov, D., Kato, K.: Anti-concentration and honest adaptive confidence bands. Ann. Stat. **42**, 1787–1818 (2014)
[9] Donoho, D.L.: Statistical estimation and optimal recovery. Ann. Stat. **22**(1), 238–270 (1994)
[10] Donoho, D.L., Liu, R.C., MacGibbon, B.: Minimax risk over hyperrectangles, and implications. Ann. Stat. **18**(3), 1416–1437 (1990)
[11] Donoho, D.L., Low, M.G.: Renormalization exponents and optimal pointwise rates of convergence. Ann. Stat. **20**(2), 944–970 (1992)
[12] Freedman, D.: On the Bernstein-von Mises theorem with infinite-dimensional parameters. Ann. Stat. **27**(4), 1119–1140 (1999)
[13] Giné, E., Nickl, R.: Confidence bands in density estimation. Ann. Stat. **38**(2), 1122–1170 (2010)
[14] Knapik, B.T., Szabó, B.T., van der Vaart, A.W., van Zanten, J.H.: Bayes procedures for adaptive inference in inverse problems for the white noise model. ArXiv e-prints (2012)
[15] Knapik, B.T., van der Vaart, A.W., van Zanten, J.H.: Bayesian inverse problems with Gaussian priors. Ann. Stat. **39**(5), 2626–2657 (2011)
[16] Nickl, R., Szabó, B.: A sharp adaptive confidence ball for self-similar functions. ArXiv e-prints (2014)
[17] Picard, D., Tribouley, K.: Adaptive confidence interval for pointwise curve estimation. Ann. Stat. **28**(1), 298–335 (2000)
[18] Ray, K.: Bernstein-von Mises theorems for adaptive Bayesian nonparametric procedures. ArXiv e-prints (2014)
[19] Rivoirard, V., Rousseau, J.: Bernstein–von Mises theorem for linear functionals of the density. Ann. Stat. **40**(3), 1489–1523 (2012)
[20] Serra, P., Krivobokova, T.: Adaptive empirical Bayesian smoothing splines. ArXiv e-prints (2014)
[21] Szabo, B.T., van der Vaart, A.W., van Zanten, J.H.: Empirical bayes scaling of gaussian priors in the white noise model. Electron. J. Stat. **7**, 991–1018 (2013)
[22] Szabo, B.T., van der Vaart, A.W., van Zanten, J.H.: Frequentist coverage of adaptive nonparametric Bayesian credible sets. Ann. Stat. (2014) http://www.imstat.org/aos/future_papers.html
[23] Szabo, B.T., van der Vaart, A.W., van Zanten, J.H.: Honest Bayesian confidence sets for the L2-norm. Stat. Plann. Inference (2014) http://www.sciencedirect.com/science/article/pii/S0378375814001244
[24] Weimin Yoo, W., Ghosal, S.: Supremum Norm Posterior Contraction and Credible Sets for Nonparametric Multivariate Regression. ArXiv e-prints (2014)

Part II
Applications and Case Studies

Chapter 9
Identifying the Infectious Period Distribution for Stochastic Epidemic Models Using the Posterior Predictive Check

Muteb Alharthi, Philip O'Neill, and Theodore Kypraios

Abstract Under the Bayesian framework, we develop a novel method for assessing the goodness of fit for the SIR (susceptible-infective-removed) stochastic epidemic model. This method seeks to determine whether or not one can identify the infectious period distribution based only on a set of partially observed data using a posterior predictive distribution approach. Our criterion for assessing the model's goodness of fit is based on the notion of Bayesian residuals.

Key words: SIR, Epidemic models, MCMC, Predictive distribution, Bayesian residual

9.1 Introduction

Poor fit of a statistical model to data can result in suspicious outcomes and misleading conclusions. Although the area of parameter estimation for stochastic epidemic models has been a subject of considerable research interest in recent years (see, e.g., [1, 7, 9]), more work is needed for the model assessment in terms of developing new methods and procedures to evaluate goodness of fit for epidemic models. Therefore, it is of importance to seek a method for assessing the quality of fitting a stochastic epidemic model to a set of epidemiological data.

The most well-known stochastic model for the transmission of infectious diseases is considered, that is the SIR (susceptible-infective-removed) stochastic epidemic model. We recall methods of Bayesian inference using Markov chain Monte Carlo (MCMC) techniques for the SIR model where partial temporal data

M. Alharthi (✉) • P. O'Neill • T. Kypraios
School of Mathematical Sciences, University of Nottingham, Nottingham, UK
e-mail: pmxma16@nottingham.ac.uk; philip.oneill@nottingham.ac.uk; heodore.kypraios@nottingham.ac.uk

S. Frühwirth-Schnatter et al. (eds.), *Bayesian Statistics from Methods to Models and Applications*, Springer Proceedings in Mathematics & Statistics 126,
DOI 10.1007/978-3-319-16238-6_9

are available. Then, a new simulation-based goodness of fit method is presented. This method explores whether or not the infectious period distribution can be identified based on removal data using a posterior predictive model checking procedure.

9.2 Model, Data and Inference

We consider a SIR stochastic epidemic model [2] in which the rate of new infections at time t is given by $\beta n^{-1}X(t)Y(t)$, where $X(t)$ and $Y(t)$ represent the number of susceptible and infective individuals at t in a closed homogeneous population of size $\mathcal{N} = n+1$, which consists of n initial susceptibles and one initial infective, and β denotes the infection rate parameter.

Following [3, 5], let $f_{T_I}(\cdot)$ denote the probability density function of T_I (the length of the infectious period, which is assumed to be a continuous random variable) and let θ indicate the parameter governing T_I. Also, define $\mathbf{I} = (I_1,\ldots,I_{n_I})$ and $\mathbf{R} = (R_1,\ldots,R_{n_R})$, where I_j and R_j are the infection and removal times of individual j and where we shall assume, for simplicity, that the total number of infections and removals are equal, that is $n_I = n_R = m$ (this assumption can be relaxed, see [8] for the details). Assuming a fully observed epidemic (complete data) with the initial infective labelled z such that $I_z < I_j$ for all $j \neq z$, the likelihood of the data given the model parameters is

$$L(\mathbf{I},\mathbf{R}|\beta,\theta,z) = \left(\prod_{j=1, j\neq z}^{m} \beta n^{-1} Y(I_j-) \right) \cdot \exp\left(-\beta n^{-1} A\right) \cdot \prod_{j=1}^{m} f_{T_I}\left(R_j - I_j\right),$$

where $A = \sum_{j=1}^{m} \sum_{k=1}^{\mathcal{N}} (R_j \wedge I_k - I_k \wedge I_j)$ with $I_k = \infty$ for $k = m+1,\ldots,\mathcal{N}$. Here, I_j- denotes the time just prior to I_j and R_j- is defined similarly.

Unfortunately, incomplete data (where we observe only removal times) are the most common type of epidemic data. As a result, the likelihood of observing only the removal times given the model parameters is intractable. One solution to make the likelihood tractable is to use the data augmentation technique by treating the missing data as extra (unknown) parameters [8]. For instance, let $T_I \sim Exp(\gamma)$, where γ is referred to as the removal rate. By adopting a Bayesian framework and assigning conjugate gamma prior distributions to the model parameters [8] that are $\beta \sim Gamma(\lambda_\beta, \nu_\beta)$, (with mean = λ_β/ν_β) and $\gamma \sim Gamma(\lambda_\gamma, \nu_\gamma)$, we get the following full conditional posterior distributions:

$$\beta|\gamma,\mathbf{I},\mathbf{R} \sim Gamma\left(\lambda_\beta + m - 1, \nu_\beta + n^{-1}A\right),$$

$$\gamma|\beta,\mathbf{I},\mathbf{R} \sim Gamma\left(\lambda_\gamma + m, \nu_\gamma + \sum_{j=1}^{m}(R_j - I_j)\right),$$

as well as

$$\pi(\mathbf{I}|\beta,\gamma,\mathbf{R}) \propto \left(\prod_{j=1,j\neq z}^{m} Y(I_j-)\right)\cdot \exp\left(-\beta n^{-1}A\right)\cdot \prod_{j=1}^{m}\exp\left(-\gamma(R_j-I_j)\right).$$

The model parameters β and γ can be updated using Gibbs sampling steps as they have closed form of the posterior distributions. However, the infection times need to be updated using a Metropolis–Hastings step. Having done that, we can obtain samples from the marginal posterior distributions of the model parameters.

When the length of the infectious periods is assumed to be constant, we have two model parameters to be updated, namely the mean of the infectious period $E(T_I)=c$ and the infection rate parameter β. However, if we let the infectious periods to have a gamma distribution $Gamma(\alpha,\delta)$, in addition to estimating the infection rate parameter β, we shall assume for computational reasons that the gamma shape parameter α is known (although it can be considered as unknown parameter to be estimated from the data, see [6] for the details) and the scale parameter δ is unknown and has to be estimated using MCMC output.

9.3 Methodology

We are concerned with identifying the infectious period distribution of the SIR model based only on removal data. In the SIR stochastic epidemic model, regardless of the type of infectious period distribution (we consider Exponential, Gamma and Constant), the total population size is constant and satisfies $\mathcal{N}=X(t)+Y(t)+Z(t)$, where $Z(t)$ denotes the number of removed individuals at event time t with $X(0)\geq 1, Y(0)\geq 1$ and $Z(0)=0$; note that $Z(s)\leq Z(t)$ for any $0\leq s\leq t; s,t\geq 0$.

However, due to the fact that epidemic data are partially observed it is sufficient for our purpose to consider only the times when removals occur instead of looking at all event times. Assuming that all infected individuals are removed by the end of the epidemic, the behaviour of the three models in terms of $Z(r_1), Z(r_2), \ldots,$ differs, where r_j represents the j-th removal time.

We turn our attention to taking advantage of this difference to distinguish between these three models when fitting them to data in the case of partial observations. Let $\mathbf{R}^{obs}=(R_1^{obs},\ldots,R_m^{obs})$ and $\mathbf{R}^{rep}=(R_1^{rep},\ldots,R_m^{rep})$ denote the observed and replicated removal times, respectively, and also let $\pi(\mathbf{R}^{rep\,i}|\mathbf{R}^{obs})$ represent the removal times predictive distribution. Then our proposed method can be generally described by the **Algorithm 1**.

Step 3 in the **Algorithm 1** can be done simply by keeping simulating (until the desired sample size is obtained) from the model using the model parameter posterior distributions while rejecting simulations that do not match the observed final size.

Algorithm 1 Generic algorithm for our method

1. Given $\mathbf{R}^{obs}$, fit an SIR model using MCMC to get samples from $\pi(\beta|\mathbf{R}^{obs})$ and $\pi(\theta|\mathbf{R}^{obs})$.
2. Draw $\beta^i \sim \pi(\beta|\mathbf{R}^{obs})$ and $\theta^i \sim \pi(\theta|\mathbf{R}^{obs})$, $i = 1,\ldots,M$.
3. Use β^i and θ^i to draw samples from $\pi(\mathbf{R}^{rep\,i}|\mathbf{R}^{obs})$ conditioning on $m^{rep\,i} = m^{obs}$.
4. Compare $\mathbf{R}^{obs}$ and $\pi(\mathbf{R}^{rep\,i}|\mathbf{R}^{obs})$ graphically as well as using Bayesian residual criterion.

9.4 Illustration

To illustrate our method, 93 removal times were simulated from an SIR model in which $T_I \sim Exp(0.5)$ and $\beta = 1.5$ in a population of size $\mathcal{N} = 100$, that consists of $n = 99$ initial susceptibles and one initial infective.

Throughout the analysis, uninformative gamma prior distributions with parameters $\lambda_\beta = \lambda_\gamma = \lambda_\delta = 1$ and $\nu_\beta = \nu_\gamma = \nu_\delta = 0.001$ were set to the parameters of the SIR models and it was assumed that the gamma shape parameter, when fitting the SIR model with gamma infectious period $T_I \sim Gamma(\alpha,\delta)$ is known ($\alpha = 10$).

By looking at Fig. 9.1, it is clearly noticeable that the observed data fit very well within the predictive distribution of the exponential SIR model, the model that has generated the data.

As mentioned above, our preferred criterion to measure the goodness of fit is the Bayesian residual [4], that is, conditioning on $m^{rep\,i} = m^{obs}$,

$$d_j = R_j^{obs} - E(R_j^{rep\,i}|\mathbf{R}^{obs}),\ j = 1,\ldots,m,$$

where $E(R_j^{rep\,i}|\mathbf{R}^{obs}) = \int R_j^{rep\,i}\,\pi(R_j^{rep\,i}|\mathbf{R}^{obs})\,dR_j^{rep\,i} \approx \frac{1}{M}\sum_{i=1}^{M} R_j^{rep\,i}$.

It is worth mentioning here that the quantity $\sum_{j=1}^{m} d_j^2$ could provide an overall measure of fit. Figure 9.2 shows the Bayesian residual distributions for the three models in which it is qualitatively obvious that there is a high density accumulated near zero, coming from the Exponential SIR model, compared to the other two models. On top of that, quantitatively, the sum of the squared Bayesian residuals $\sum_{j=1}^{m} d_j^2$ are 96.3, 354.7 and 812.6 for the Exponential, Gamma and Constant SIR models, respectively. Therefore, as expected, the Exponential SIR model, from which the data was generated, has the smallest value of the sum of the squared Bayesian residuals.

9.5 Conclusion

Bayesian inference for the SIR model has been introduced, where the epidemic outbreak is partially observed. We have proposed a method to assess the goodness of fit for the SIR stochastic model based only on removal data. A simulation study

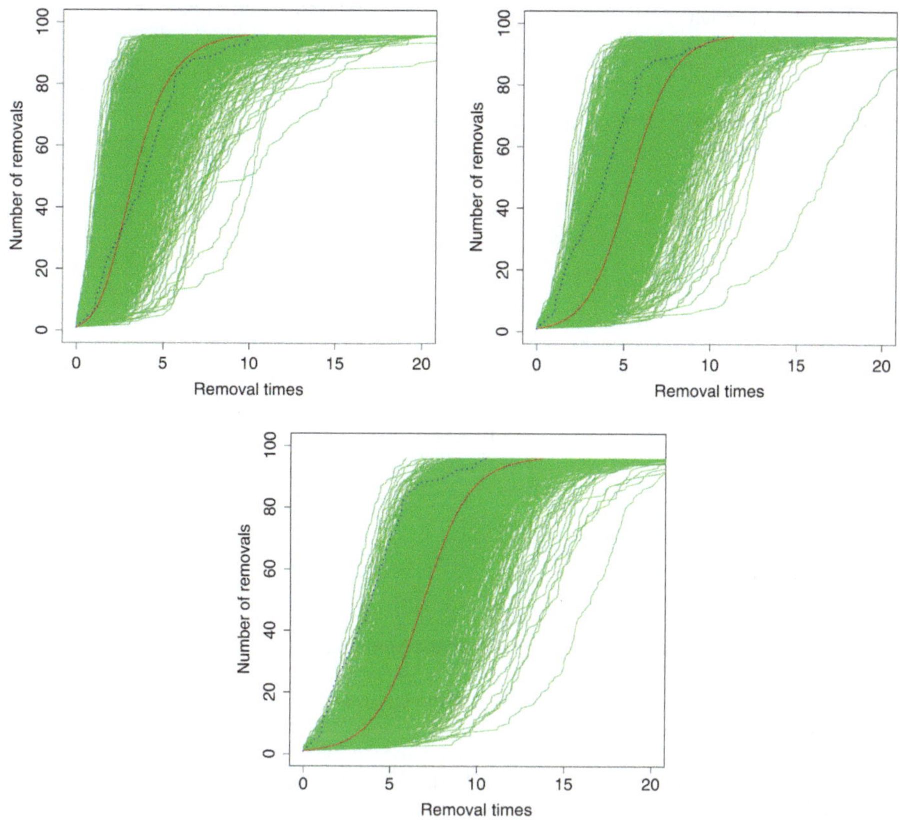

Fig. 9.1 Comparison of the removal times predictive distribution for the three SIR models (*top left*: Exponential, *top right*: Gamma, *bottom*: Constant) based on 1,000 realizations and conditioning on the observed final size, where the *dotted line* indicates the observed data and the *solid line* represents the predictive mean

has been performed to test the proposed method. Using the posterior predictive assessment for checking models, this diagnostic method is able to identify the true model reasonably well.

One advantage of this method is that it looks explicitly at the discrepancy between observed and predicted data, which avoids using unobserved quantities in the process of assessment, see [10] as an example. Furthermore, this method is still valid when including an extra state, the exposed period, to the SIR model in which individuals in this state are infected but not yet infectious.

Acknowledgements The first author is supported by a scholarship from Taif University, Taif, Saudi Arabia.

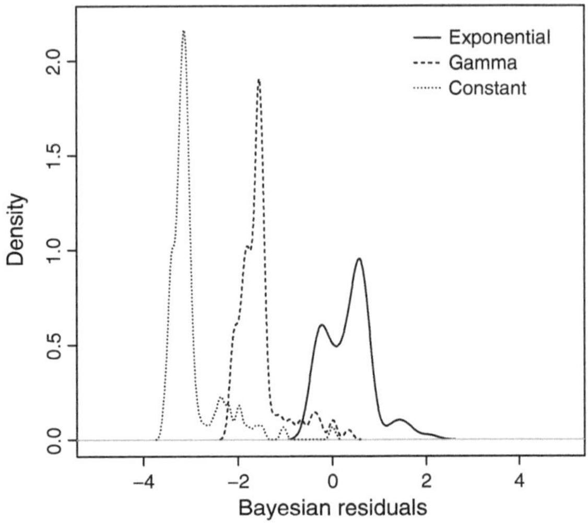

Fig. 9.2 The Bayesian residual distributions for each SIR model based on 1,000 samples from the conditioning predictive distribution for the three models

References

[1] Andersson, H., Britton, T.: Stochastic Epidemic Models and Their Statistical Analysis, vol. 4. Springer, New York (2000)

[2] Bailey, N.T.J.: The Mathematical Theory of Infectious Diseases and Its Applications. Charles Griffin & Company, London (1975)

[3] Britton, T., O'Neill, P.D.: Bayesian inference for stochastic epidemics in populations with random social structure. Scand. J. Stat. **29**(3), 375–390 (2002)

[4] Gelfand, A.E.: Model determination using sampling-based methods. In: Gilks, W.R., Richardson, S., Spiegelhalter, D.J. (eds.) Markov Chain Monte Carlo in Practice, pp. 145–161. Springer, New York (1996)

[5] Kypraios, T.: Efficient Bayesian inference for partially observed stochastic epidemics and a new class of semi-parametric time series models. Ph.D. thesis, Lancaster University (2007)

[6] Neal, P., Roberts, G.O.: A case study in non-centering for data augmentation: stochastic epidemics. Stat. Comput. **15**(4), 315–327 (2005)

[7] O'Neill, P.D.: Introduction and snapshot review: relating infectious disease transmission models to data. Stat. Med. **29**(20), 2069–2077 (2010)

[8] O'Neill, P.D., Roberts, G.O.: Bayesian inference for partially observed stochastic epidemics. J. Roy. Stat. Soc. Ser. A (Stat. Soc.) **162**(1), 121–129 (1999)

[9] Streftaris, G., Gibson, G.J.: Bayesian inference for stochastic epidemics in closed populations. Stat. Model. **4**(1), 63–75 (2004)

[10] Streftaris, G., Gibson, G.J.: Non-exponential tolerance to infection in epidemic systems–modeling, inference, and assessment. Biostatistics **13**(4), 580–593 (2012)

Chapter 10
A New Strategy for Testing Cosmology with Simulations

Madhura Killedar, Stefano Borgani, Dunja Fabjan, Klaus Dolag, Gian Luigi Granato, Susana Planelles, and Cinthia Ragone-Figueroa

Abstract Structural properties of clusters of galaxies have been routinely used to claim either tension or consistency with the fiducial cosmological hypothesis, known as ΛCDM. However, standard approaches are unable to quantify the preference for one hypothesis over another. We advocate using a 'weighted' variant of approximate Bayesian computation (ABC), whereby the parameters of the strong lensing-mass scaling relation, α and β, are treated as the summary statistics. We demonstrate then, for the first time, the procedure for estimating the likelihood for observing α and β under the ΛCDM framework. We employ computer simulations for producing mock samples, and account for variation between samples for modelling the likelihood function. We also consider the effects on the likelihood, and consequential ability to compare competing hypotheses, if only simplistic computer simulations are available.

Key words: Bayesian statistics, ABC, Simulations, Cosmology, Galaxy clusters

---

M. Killedar (✉)
Universitäts-Sternwarte München, Ludwig Maximilians University, München, Germany
e-mail: killedar@usm.lmu.de

S. Borgani
Dipartimento di Fisica, Università di Trieste, Trieste, Italy
e-mail: borgani@oats.inaf.it

D. Fabjan
SPACE-SI, Slovenian Centre of Excellence for Space Sciences and Technologies, Aškerčeva 12, 1000 Ljubljana, Slovenia

K. Dolag
Universitaets-Sternwarte Muenchen, Scheinerstrasse 1, D-81679, Muenchen, Germany

G.L. Granato • S. Planelles
Dipartimento di Fisica dell'Universita di Trieste, Sezione di Astronomia, Via Tiepolo 11, I-34131 Trieste, Italy

C. Ragone-Figueroa
Instituto de Astronomia Teorica y Experimental (IATE), Consejo Nacional de Investigaciones Cientýficas y Tecnicas de la Republica Argentina (CONICET), Observatorio Astronomico, Universidad Nacional de Cordoba, Laprida 854, X5000BGR Cordoba, Argentina

S. Frühwirth-Schnatter et al. (eds.), *Bayesian Statistics from Methods to Models and Applications*, Springer Proceedings in Mathematics & Statistics 126,
DOI 10.1007/978-3-319-16238-6_10

10.1 Introduction

Our standard model of cosmology, ΛCDM, is one in which our universe is made up primarily of dark energy, has a large amount of dark matter, and only a small fraction of ordinary matter; it is currently undergoing a period of expansion and an expansion that is accelerating. This model appears to describe the contents and evolution of the universe very well, and has been determined through the analysis of several astrophysical objects and phenomena. One additional dataset with the potential to provide complementary information is the mass-structure of clusters of galaxies [1, 2, 22, 24]. These objects contain hundreds to thousands of galaxies, as well as hot gas and dark matter, and they gravitationally lens and distort the images of more distant galaxies.

Strong lensing efficiencies, as characterised by the effective Einstein radii (denoted θ_E) scale well with the mass of clusters at large over-densities [14]. If any given set of galaxy clusters sample are, in fact, stronger lenses than predicted by the ΛCDM model, they will have larger θ_E for a given total mass at low over-densities (e.g., M_{500}). The earliest studies of similar galaxy-cluster properties revealed a significant difference between the observations and ΛCDM predictions [1, 15]. Thus began the hunt for solutions to the so-called tension with ΛCDM cosmology.

Previous works in the literature have claimed either 'tension' or 'consistency' with ΛCDM, or insufficient data [6, 11, 18, 21, 23, 27], but do not allow one to compare competing cosmological hypotheses. In the present work, we propose a Bayesian approach to the cosmological test using galaxy cluster strong lensing properties.

10.2 The Bayesian Framework

A Bayesian approach is advocated, in which one determines the relative preference of two hypothetical cosmological models, C_1 and C_2, in light of the data D, by calculating the Bayes factor B:

$$B = \frac{\mathscr{L}(D|C_1)}{\mathscr{L}(D|C_2)} \tag{10.1}$$

where $\mathscr{L}$ denotes the likelihood of the data assuming a cosmology.

The aim then is to calculate, under one chosen hypothesis: ΛCDM, the likelihood of observing the structural properties of a particular sample of galaxy clusters. This sample is detected using a well-defined selection criteria and all relevant properties have been measured [5, 7, 13, 16, 17, 26, 27].

10.2.1 Weighted ABC

Achieving the aforementioned goal is non-trivial, because a likelihood function related to the original observables (θ_E and M_{500}) is intractable. This is because: (a) computer simulations are deemed necessary for describing the irregular structure of galaxy clusters, which undergo non-linear structure formation; (b) there are a finite (and relatively small) number of clusters that can be simulated in a reasonable amount of time, and thus the full θ_E–M_{500} space cannot be sampled. Therefore, this problem is an ideal case for which one may apply a variant of approximate Bayesian computation (ABC) [4, 25]. What we propose is not a likelihood-free approach, however, and rather than rejecting mock samples that are dissimilar to the real data, they are down-weighted. Thus, we refer to the novel approach described below as Weighted ABC.

We assume a power-law relation between the strong lensing and mass proxies, and perform a fitting to the following function in logarithmic space[1]:

$$\log\left[\frac{M_{500}}{9\times10^{14}M_{\odot}}\right]=\alpha\log\left[\frac{\theta_E}{20''}\sqrt{\frac{D_d}{D_{ds}}}\right]+\beta \qquad (10.2)$$

with parameters α and β, and aim to find the likelihood of observing *the scaling relationship*. α and β act as summary statistics for the dataset. However, rather than calculating precise values for α and β, one would determine a probability distribution that reflects the degree of belief in their respective values. The relevant fitting procedure is described in Sect. 10.2.2.

Next, we outline how to calculate the likelihood of observing α and β. In the following, ι represents background information such as knowledge of the cluster selection criteria, the method of characterising the Einstein radius, and the assumption that there exists a power-law relation between strong lensing and mass.

1. Computer simulations (see [3, 14, 19, 20]) are run within the framework of a chosen cosmological hypothesis, C. In our case, C represents the assumption that ΛCDM (or specific values for cosmological parameters) is the true description of cosmology.
2. Simulated galaxy clusters are selected according to specified criteria, ideally reflecting the criteria used to select the real clusters.
3. Different on-sky projections of these three-dimensional objects produce different apparent measurements of structural properties. Therefore, we construct a large

[1] The pivot mass $9\times10^{14}M_{\odot}$ is chosen to approximate the logarithmic average of the observed and simulated clusters. Similarly, the pivot Einstein radius is chosen to be 20 arcseconds. D_d represents the angular diameter distance from an observer on Earth to the galaxy cluster lens, while D_{ds} represents the angular diameter distance from the galaxy cluster lens to a more distant galaxy, in our case chosen to be fixed to a redshift of $z=2$.

number of mock samples from these by randomly choosing an orientation-angle for each cluster. Equation (10.2) is fit to each mock sample (See Sect. 10.2.2), to determine a posterior over α and β: $P_i(\alpha,\beta|C,\iota)$ denotes the result for the ith of N mock samples. We combine these, to give the probability, $P(\alpha,\beta|C,\iota) \equiv \sum_{i=1}^{N} P_i(\alpha,\beta|C,\iota)$, that one would observe the scaling relation $\{\alpha,\beta\}$ under the hypothesis C. The result can be interpreted as a likelihood function *as a function of data*: α and β.
4. Fit Eq. (10.2) to the data to obtain the posterior probability distribution for α and β, $P(\alpha,\beta|\iota)$. The normalised posterior is then interpreted as a single 'data point': the distribution represents the uncertainty on the measurement of α and β.
5. Calculate the likelihood, $\mathscr{L}$, of observing the α–β fit as we did, by integrating over the product of the two aforementioned posteriors—now re-labelled 'data-point' and 'likelihood function'.

The result of integrating the product of $P(\alpha,\beta|C,\iota)$ and $P(\alpha,\beta|\iota)$ for the dataset is mathematically equivalent to integrating the product for each mock separately, then taking the average over all mock samples:

$$\int \left[\frac{1}{N}\sum_{i=1}^{N} P_i(\alpha,\beta|C,\iota)\right] P(\alpha,\beta|\iota)\,\mathrm{d}\alpha\,\mathrm{d}\beta = \frac{1}{N}\sum_{i=1}^{N}\int P_i(\alpha,\beta|C,\iota)P(\alpha,\beta|\iota)\,\mathrm{d}\alpha\,\mathrm{d}\beta \tag{10.3}$$

Thus, what we have described above is equivalent to the weighting of each mock sample according to its similarity to the real data, where the metric is the convolution of the two (mock and real) posterior probability distributions $P(\alpha,\beta|\iota)$.

10.2.2 *Summary Statistic Fitting*

The summary statistics α and β are parameters of the scaling relation between strong lensing efficiency and total cluster mass [Eq. (10.2)]. The procedure for calculating $\mathscr{L}$, as described in Sect. 10.2.1, requires one to fit real or mock data to determine the posterior distribution on α and β. We employ the Bayesian linear regression method outlined in [10]. Additionally, we acknowledge that intrinsic scatter is likely to be present, and thus introduce a nuisance parameter, V, which represents intrinsic Gaussian variance orthogonal to the line.

For this subsection, we change notation in order to reduce the subscripts: the mass of the i-th cluster lens as M_i, and the scaled Einstein radius as E_i. Each data-point is denoted by the vector $\mathbf{Z}_i = [\log M_i, \log E_i]$. Their respective uncertainties (*on the logarithms*) are denoted σ_M^2 and σ_E^2. Since we assume the uncertainties for Einstein radii and cluster mass are uncorrelated, the covariance matrix, $\mathbf{S}_i$, reduces to:

$$\mathbf{S}_i \equiv \begin{pmatrix} \sigma_M^2 & 0 \\ 0 & \sigma_E^2 \end{pmatrix} \tag{10.4}$$

In the case of a mock sample of simulated clusters, $\mathbf{S}_i = 0$.

Consider now the following quantities: $\varphi \equiv \arctan\alpha$, which denotes the angle between the line and the x-axis, and $b_{\perp} \equiv \beta\cos\varphi$ which is the orthogonal distance of the line to the origin. The orthogonal distance of each data-point to the line is:

$$\Delta_i = \hat{\mathbf{v}}^{\top}\mathbf{Z}_i - \beta\cos\varphi \tag{10.5}$$

where $\hat{\mathbf{v}} = [-\sin\varphi, \cos\varphi]$ is a vector orthogonal to the line.

Therefore, the orthogonal variance is

$$\Sigma_i^2 = \hat{\mathbf{v}}^{\top}\mathbf{S}_i\hat{\mathbf{v}}. \tag{10.6}$$

Following [10], we calculate the likelihood over the three-dimensional parameter space: $\Theta_1 \equiv \{\alpha, \beta, V\}$:

$$\ln\mathscr{L} = \mathrm{K} - \sum_{i=1}^{N}\frac{1}{2}\ln(\Sigma_i^2 + V) - \sum_{i=1}^{N}\frac{\Delta_i^2}{2\Sigma_i^2 + V} \tag{10.7}$$

where K is an arbitrary constant, and the summation is over all clusters in the considered sample.

While we ultimately (aim to) provide the parameter constraints on α and β, flat priors for these tend to unfairly favour large slopes. A more sensible choice is flat for the alternative parameters φ and $b_{\perp}$. We apply a modified Jeffreys prior on V:

$$\pi(V) \propto \frac{1}{V + V_t} \tag{10.8}$$

This is linearly uniform on V for small values and logarithmically uniform on V for larger values with a turnover, V_t, chosen to reflect the typical uncertainties.

Thus, for each Θ_1, we may define an alternative set of parameters $\Theta_2 \equiv \{\varphi, b_{\perp}, V\}$, for which the prior is given by:

$$\begin{aligned}\pi(\Theta_2) &= \pi(\varphi, b_{\perp})\pi(V)\\ &\propto \pi(V)\end{aligned} \tag{10.9}$$

where $\pi(V)$ is given by Eq. 10.8. The prior on Θ_1 is then dependent on the magnitude of the Jacobian of the mapping between the two sets of parameters:

$$\begin{aligned}\pi(\Theta_1) &= \pi(\Theta_2)\det\tfrac{\partial\Theta_2}{\partial\Theta_1}\\ &\equiv \pi(\Theta_2)\tfrac{1}{(1+\alpha^2)^{3/2}}\end{aligned} \tag{10.10}$$

Boundaries on the priors are sufficiently large[2]: $-8 \le \beta \le 8$; $-40 \le \alpha \le 40$; $0 \le V \le V_{\max}$. $V_{\max}$ is chosen to reflect the overall scatter in the data. The posterior is calculated following Bayes' theorem:

$$P(\Theta_1|D) \propto \mathscr{L}(D|\Theta_1)\,\pi(\Theta_1) \tag{10.11}$$

and is normalised. In practice, the posterior distribution was sampled by employing `emcee` [8], the python implementation of the affine-invariant ensemble sampler for Markov chain Monte Carlo (MCMC) proposed by [9].

As we are interested in the constraints on α and β, we then marginalise over the nuisance parameter, V.

10.3 Results

In Fig. 10.1, we show the relation between the Einstein radii and the cluster mass M_{500}. The real cluster sample is represented by red circles. For simulated clusters, the situation is more complicated. Since different lines of sight provide a large variation in projected mass distribution, each cluster cannot be associated with an individual Einstein radius, nor a simple Gaussian or log-normal distribution [14]. We therefore measure the Einstein radius for 80 different lines of sight and, for ease of visualisation, describe the distribution of Einstein radii for each simulated cluster by a box-plot.

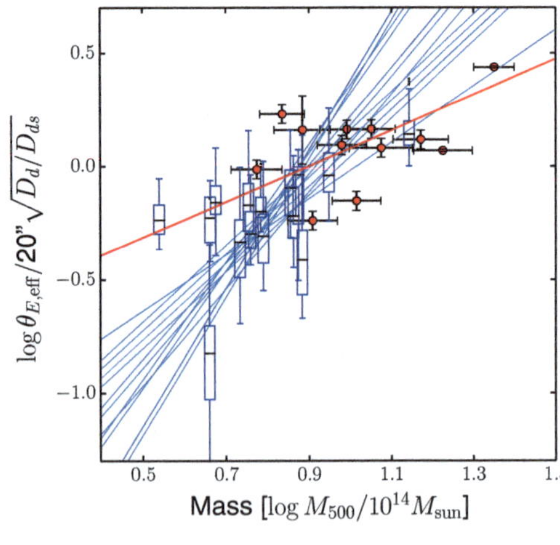

Fig. 10.1 Strong lensing efficiency, characterised by scaled Einstein radii, $\theta_{\mathrm{E,eff}}$, plotted as a function of mass. The range of Einstein radii for simulated clusters are shown by the *blue box-plots*. The *red circles* represent the real clusters. The *red line* marks the maximum a-posteriori fit to observational data, while the *thin blue lines* mark the fit to 20 randomly chosen mock samples from simulations

[2]The physically motivated choice of restricting $\alpha \ge 0$ is also explored; however, this has very minor effects on the final results despite removing the (small) secondary peak in the marginal posterior on α and β.

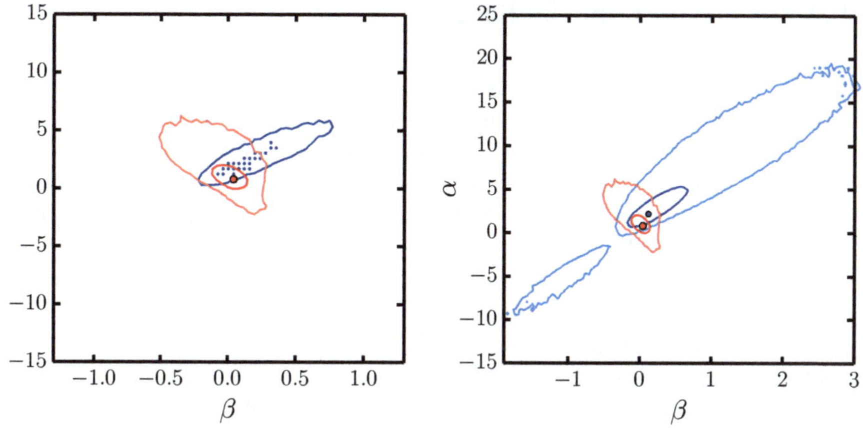

Fig. 10.2 *Left*: 1-σ and 2-σ constraints on parameters of the strong lensing—mass relation given the real cluster data (*red contours*), with a maximum a posteriori fit marked by a *red circle*. Overplotted in *blue dots* are the best fits to 80 mock observations of simulated galaxy clusters. A typical 1-σ error is shown as a *blue ellipse*. *Right*: Same as the middle panel, but the *blue circle* and *curves* mark, respectively, the maximum and the 1-σ and 2-σ contours of the likelihood function found by combining all 80 mocks. Ultimately, the likelihood, $\mathscr{L} \approx 0.3$, is found by convolving the functions marked by the *red* and *blue contours*

We fit the observational data to the lensing-mass relation and after marginalising out the nuisance parameter, V, present the posterior distribution for α and β, denoted by red contours in the left-hand panel of Fig. 10.2. This fit is reinterpreted as a single 'data-point'. To estimate the likelihood, *as a function of possible data*, we employ simulations. Many mock samples are individually fit to the lensing-mass relations; the maximum of the posterior is shown as a blue point and a typical 1-σ error shown as a blue ellipse. By adding the posteriors for each mock sample and renormalising, we estimate the required likelihood function, shown by the blue contours in the right-hand panel of Fig. 10.2. By multiplying by the 'data-point' distribution and integrating over the parameter space, we find $\mathscr{L} \approx 0.3$.

Note that one cannot comment on whether the likelihood is *large* or *small*. Currently, such simulations are only available for the fiducial ΛCDM cosmological model. However, if the same process is repeated for simulations under a different model, then the Bayes factor can be calculated [see Eq. (10.1)] and, after accounting for priors, may (or may not) reveal a preference for one of the cosmologies, in light of this data. Alternative cosmological models may include, for example, those with a different relative dark matter to dark energy ratio, interactions between the two dark components, or a different normalisation for the structure power spectrum.

10.4 Computational Challenge

The approach described above is an exciting new strategy for calculating the likelihood for observing strong lensing galaxy clusters for a chosen cosmological hypothesis. *However*, we recognise that the calculation involves running computer simulations that can take months. Computationally 'cheaper' simulations ignore several astrophysical processes in the formation of galaxy clusters and it is debatable whether these would be sufficient.

In order to determine the severity of this problem, we repeat the aforementioned procedure using galaxy cluster counterparts from such simulations, at varying levels of complexity and realism, and find that the likelihood, $\mathscr{L}$, can then vary by a factor of three or four. If the cheaper simulations are employed, then the selection criteria must also be replaced with an alternative compromise. We test this alternative and find that $\mathscr{L}$ changes by a factor of two.

Our findings suggest that if a model-comparison study was carried out using a simulation based on an alternative cosmological hypothesis and resulting in a Bayes factor of 20 or more [see Eq. (10.1)], then the cheaper simulations (or toy models based on these) would be sufficient. However, in the event that the Bayes factor B is found to be smaller, then the computationally expensive but realistic simulations would be necessary.

Acknowledgements The authors thank Ewan Cameron, Geraint Lewis, Jiayi Liu, Hareth Saad Mahdi, Julio Navarro, Matthias Redlich, Christian Robert, Alexandro Saro, Anja von der Linden and Adi Zitrin for helpful discussions, and Giuseppe Murante for his involvement in running the simulations.

This work has made use of the following open-source software: python,[3] numpy,[4] scipy,[5] matplotlib [12] and cosmolopy.[6]

MK acknowledges a fellowship from the European Commission's Framework Programme 7, through the Marie Curie Initial Training Network CosmoComp (PITN-GA-2009-238356) and support by the DFG project DO 1310/4-1. SP acknowledges support by the PRIN-INAF09 project "Towards an Italian Network for Computational Cosmology" and by the PRIN-MIUR09 "Tracing the growth of structures in the Universe" as well as partial support by *Spanish Ministerio de Ciencia e Innovación* (MICINN) (grants AYA2010-21322-C03-02 and CONSOLIDER2007-00050). Simulations have been carried out at the CINECA supercomputing Centre in Bologna (Italy), with CPU time assigned through ISCRA proposals and through an agreement with University of Trieste. This work has been supported by the PRIN-MIUR 2012 grant "Evolution of Cosmic Baryons", by the PRIN-INAF 2012 grant "Multi-scale Simulations of Cosmic Structure" and by the INFN "INDARK" grant. We acknowledge support from "Consorzio per la Fisica - Trieste".

[3]http://www.python.org.

[4]http://www.numpy.org.

[5]http://www.scipy.org.

[6]http://roban.github.com/CosmoloPy/.

References

[1] Bartelmann, M., Huss, A., Colberg, J.M., Jenkins, A., Pearce, F.R.: Arc statistics with realistic cluster potentials. IV. Clusters in different cosmologies. A&A **330**, 1–9 (1998)

[2] Bartelmann, M., Meneghetti, M., Perrotta, F., Baccigalupi, C., Moscardini, L.: Arc statistics in cosmological models with dark energy. A&A **409**, 449–457 (2003). Doi:10.1051/0004-6361:20031158

[3] Bonafede, A., Dolag, K., Stasyszyn, F., Murante, G., Borgani, S.: A non-ideal MHD gadget: simulating massive galaxy clusters. ArXiv e-prints (2011)

[4] Cameron, E., Pettitt, A.N.: Approximate Bayesian computation for astronomical model analysis: a case study in galaxy demographics and morphological transformation at high redshift. MNRAS **425**, 44–65 (2012). Doi:10.1111/j.1365-2966.2012.21371.x

[5] Coe, D., Zitrin, A., Carrasco, M., Shu, X., Zheng, W., Postman, M., Bradley, L., Koekemoer, A., Bouwens, R., Broadhurst, T., Monna, A., Host, O., Moustakas, L.A., Ford, H., Moustakas, J., van der Wel, A., Donahue, M., Rodney, S.A., Benítez, N., Jouvel, S., Seitz, S., Kelson, D.D., Rosati, P.: CLASH: three strongly lensed images of a candidate z $\approx$ 11 galaxy. ApJ **762**, 32 (2013). Doi:10.1088/0004-637X/762/1/32

[6] Dalal, N., Holder, G., Hennawi, J.F.: Statistics of giant arcs in galaxy clusters. ApJ **609**, 50–60 (2004). Doi:10.1086/420960

[7] Ebeling, H., Barrett, E., Donovan, D., Ma, C.J., Edge, A.C., van Speybroeck, L.: A complete sample of 12 very X-ray luminous galaxy clusters at $z > 0.5$. ApJ Lett. **661**, L33–L36 (2007). Doi:10.1086/518603

[8] Foreman-Mackey, D., Hogg, D.W., Lang, D., Goodman, J.: emcee: the MCMC hammer. PASP **125**, 306–312 (2013). Doi:10.1086/670067

[9] Goodman, J., Weare, J.: Ensemble samplers with affine invariance. App. Math. Comput. Sci. **5**(1), 65–80 (2010)

[10] Hogg, D.W., Bovy, J., Lang, D.: Data analysis recipes: fitting a model to data. ArXiv e-prints (2010)

[11] Horesh, A., Maoz, D., Hilbert, S., Bartelmann, M.: Lensed arc statistics: comparison of millennium simulation galaxy clusters to hubble space telescope observations of an X-ray selected sample. MNRAS **418**, 54–63 (2011). Doi:10.1111/j.1365-2966.2011.19293.x

[12] Hunter, J.D.: Matplotlib: a 2d graphics environment. Comput. Sci. Eng. **9**(3), 90–95 (2007)

[13] Johnson, T.L., Sharon, K., Bayliss, M.B., Gladders, M.D., Coe, D., Ebeling, H.: Lens models and magnification maps of the six hubble Frontier fields clusters. ArXiv e-prints (2014)

[14] Killedar, M., Borgani, S., Meneghetti, M., Dolag, K., Fabjan, D., Tornatore, L.: How baryonic processes affect strong lensing properties of simulated galaxy clusters. MNRAS **427**, 533–549 (2012). Doi:10.1111/j.1365-2966.2012.21983.x

[15] Li, G.L., Mao, S., Jing, Y.P., Bartelmann, M., Kang, X., Meneghetti, M.: Is the number of giant arcs in ΛCDM consistent with observations? ApJ **635**, 795–805 (2005). Doi:10.1086/497583

[16] Mantz, A., Allen, S.W., Ebeling, H., Rapetti, D., Drlica-Wagner, A.: The observed growth of massive galaxy clusters - II. X-ray scaling relations. MNRAS **406**, 1773–1795 (2010). Doi:10.1111/j.1365-2966.2010.16993.x

[17] Medezinski, E., Umetsu, K., Nonino, M., Merten, J., Zitrin, A., Broadhurst, T., Donahue, M., Sayers, J., Waizmann, J.C., Koekemoer, A., Coe, D., Molino, A., Melchior, P., Mroczkowski, T., Czakon, N., Postman, M., Meneghetti, M., Lemze, D., Ford, H., Grillo, C., Kelson, D., Bradley, L., Moustakas, J., Bartelmann, M., Benítez, N., Biviano, A., Bouwens, R., Golwala, S., Graves, G., Infante, L., Jiménez-Teja, Y., Jouvel, S., Lahav, O., Moustakas, L., Ogaz, S., Rosati, P., Seitz, S., Zheng, W.: CLASH: complete lensing analysis of the largest cosmic lens MACS J0717.5+3745 and surrounding structures. ApJ **777**, 43 (2013). Doi:10.1088/0004-637X/777/1/43

[18] Meneghetti, M., Fedeli, C., Zitrin, A., Bartelmann, M., Broadhurst, T., Gottlöber, S., Moscardini, L., Yepes, G.: Comparison of an X-ray-selected sample of massive lensing clusters with the MareNostrum Universe ΛCDM simulation. A&A **530**, A17+ (2011). Doi:10.1051/0004-6361/201016040

[19] Planelles, S., Borgani, S., Fabjan, D., Killedar, M., Murante, G., Granato, G.L., Ragone-Figueroa, C., Dolag, K.: On the role of AGN feedback on the thermal and chemo-dynamical properties of the hot intracluster medium. MNRAS **438**, 195–216 (2014). Doi:10.1093/mnras/stt2141

[20] Ragone-Figueroa, C., Granato, G.L., Murante, G., Borgani, S., Cui, W.: Brightest cluster galaxies in cosmological simulations: achievements and limitations of active galactic nuclei feedback models. MNRAS **436**, 1750–1764 (2013). Doi:10.1093/mnras/stt1693

[21] Sereno, M., Giocoli, C., Ettori, S., Moscardini, L.: The mass-concentration relation in lensing clusters: the role of statistical biases and selection effects. ArXiv e-prints (2014)

[22] Takahashi, R., Chiba, T.: Gravitational lens statistics and the density profile of Dark Halos. ApJ **563**, 489–496 (2001). Doi:10.1086/323961

[23] Waizmann, J.C., Redlich, M., Meneghetti, M., Bartelmann, M.: The strongest gravitational lenses: III. The order statistics of the largest Einstein radii. ArXiv e-prints (2014)

[24] Wambsganss, J., Bode, P., Ostriker, J.P.: Giant arc statistics in concord with a concordance lambda cold dark matter universe. ApJ Lett. **606**, L93–L96 (2004). Doi:10.1086/421459

[25] Weyant, A., Schafer, C., Wood-Vasey, W.M.: Likelihood-free cosmological inference with type ia supernovae: approximate bayesian computation for a complete treatment of uncertainty. ApJ **764**, 116 (2013). Doi:10.1088/0004-637X/764/2/116

[26] Zheng, W., Postman, M., Zitrin, A., Moustakas, J., Shu, X., Jouvel, S., Høst, O., Molino, A., Bradley, L., Coe, D., Moustakas, L.A., Carrasco, M., Ford, H., Benítez, N., Lauer, T.R., Seitz, S., Bouwens, R., Koekemoer, A., Medezinski, E., Bartelmann, M., Broadhurst, T., Donahue, M., Grillo, C., Infante, L., Jha, S.W., Kelson, D.D., Lahav, O., Lemze, D., Melchior, P., Meneghetti, M., Merten, J., Nonino, M., Ogaz, S., Rosati, P., Umetsu, K., van der Wel, A.: A magnified young galaxy from about 500 million years after the Big Bang. Nature **489**, 406–408 (2012). Doi:10.1038/nature11446

[27] Zitrin, A., Broadhurst, T., Barkana, R., Rephaeli, Y., Benítez, N.: Strong-lensing analysis of a complete sample of 12 MACS clusters at $z > 0.5$: mass models and Einstein radii. MNRAS **410**, 1939–1956 (2011). Doi:10.1111/j.1365-2966.2010.17574.x

Chapter 11
Formal and Heuristic Model Averaging Methods for Predicting the US Unemployment Rate

Jeremy Kolly

Abstract We consider a logistic transform of the monthly US unemployment rate. For this time series, a pseudo out-of-sample forecasting competition is held between linear and nonlinear models and averages of these models. To combine predictive densities, we use two complementary methods: Bayesian model averaging and optimal pooling. We select the individual models combined by these methods with the evolution of Bayes factors over time. Model estimation is carried out using Markov chain Monte Carlo algorithms and predictive densities are evaluated with statistical tests and log scores. The sophisticated averages of linear and nonlinear models turn out to be valuable tools for predicting the US unemployment rate in the short-term.

Key words: Nonlinearity, Model combination, Markov chain Monte Carlo methods, Bayes factors, Forecast evaluation

11.1 Introduction

Many studies point out that nonlinear models are able to yield superior predictions of the US unemployment rate [1, 2, 4, 8, 11, 12]. Among them, [2, 4] argue in favor of the logistic smooth transition autoregression (LSTAR). This nonlinear regime-switching model, proposed by [13], can be written as:

$$y_t = \phi_{10} + \sum_{j=1}^{p} \phi_{1j} y_{t-j} + G(s_t;\gamma,c)\left(\phi_{20} + \sum_{j=1}^{p} \phi_{2j} y_{t-j}\right) + \varepsilon_t,$$

$$G(s_t;\gamma,c) = \frac{1}{1+\exp[-\gamma^2(s_t - c)]},$$

J. Kolly (✉)
Finance, Insurance and Real Estate Department, Laval University, Quebec, Quebec, Canada

Department of Management, Fribourg University, Fribourg, Switzerland
e-mail: jeremy.kolly.1@ulaval.ca

S. Frühwirth-Schnatter et al. (eds.), *Bayesian Statistics from Methods to Models and Applications*, Springer Proceedings in Mathematics & Statistics 126,
DOI 10.1007/978-3-319-16238-6_11

where the ε_t are i.i.d. $N(0,\sigma^2)$ and where the logistic transition function $G(\cdot)$ depends on the observable transition variable s_t and contains γ and c; the smoothness and location parameters, respectively.

In the contributions mentioned previously, the linear models are found to be good competitors. This may mean that linear and nonlinear models provide complementary descriptions of the US unemployment process. The present research takes this possibility into account by investigating the predictive performance of averages of linear and nonlinear models. Some of the above-mentioned studies consider model combination. However, their approaches are either limited or different. Furthermore, note that we will combine Bayesian predictive densities. They have the advantage of being small-sample results that incorporate parameter uncertainty.

The plan of this chapter is the following. Section 11.2 presents the model averaging methods. Section 11.3 describes the forecasting experiment. Section 11.4 shows real-time weights over the forecasting period and evaluates predictive performance. Section 11.5 concludes.

11.2 Model Averaging Methods

Consider the model space $\mathcal{M} = \{M_1,\dots,M_K\}$ where each model delivers a predictive density $p(y_{T+1}|y_{1:T},M_k)$ for the future observation y_{T+1} given the sample $y_{1:T} = (y_1,\dots,y_T)'$. These predictive densities can be used to form the mixture density:

$$p_{w_T}(y_{T+1}|y_{1:T}) = \sum_{k=1}^{K} w_{T,k}\, p(y_{T+1}|y_{1:T},M_k), \tag{11.1}$$

where the weight vector $w_T = (w_{T,1},\dots,w_{T,K})'$ depends on data until time T and satisfies $\sum_{k=1}^{K} w_{T,k} = 1$ and $w_{T,1},\dots,w_{T,K} \geq 0$. The naive equally weighted model averaging (EWMA) method results when $w_{T,k} = 1/K$ for all k. By setting $w_{T,k} = p(M_k|y_{1:T})$ for all k, we obtain the formal Bayesian model averaging (BMA) method proposed by [9]. Assuming equal prior model probabilities, the kth posterior model probability (PMP) can be written as:

$$p(M_k|y_{1:T}) = \frac{p(y_{1:T}|M_k)}{\sum_{l=1}^{K} p(y_{1:T}|M_l)}.$$

In what follows, marginal likelihoods $p(y_{1:T}|M_k)$ are estimated by bridge sampling [10].

BMA presumes that the data generating process (DGP) belongs to $\mathcal{M}$. As this is questionable, we also consider a heuristic method that does not make this assumption; the optimal pooling (OP) method developed by [5, 6]. The OP weights are obtained by solving:

$$\max_{w_T} \sum_{t=t_0+1}^{T} \ln\left[\sum_{k=1}^{K} w_{T,k}\, p(y_t|y_{1:t-1}, M_k)\right]$$

$$\text{subject to} \sum_{k=1}^{K} w_{T,k} = 1 \text{ and } w_{T,1}, \ldots, w_{T,K} \geq 0,$$

where the objective function is the cumulative log score of (11.1) over $y_{t_0+1:T}$ given the training sample $y_{1:t_0}$.

11.3 Setting Up the Experiment

Denote by u_t the seasonally adjusted monthly US unemployment rate in percentage points from 1:1948 for civilians of 20 years and over. A forecasting competition is held between autoregressive (AR), LSTAR and random walk (RW) models for $y_t = \ln[0.01u_t/(1-0.01u_t)]$ and averages of these models. For the LSTAR model, we set $s_t = u_{t-1} - u_{t-13}$ as recommended in [2] where several possible definitions of s_t are compared with Bayes factors on a similar data set. Our prior choices are summarized in Table 11.1. We estimate the AR model with the Gibbs sampler, the LSTAR model with the Metropolis-within-Gibbs developed in [2] and use analytical results for the RW model. The composition of the model averages in competition is determined using Fig. 11.1; the four emerging models are retained for the BMA, OP, and EWMA methods. Moreover, as OP may sometimes attribute positive weights to inferior models, the RW model is also retained for this method. Finally, one-month ahead predictive densities of individual models and model combinations are simulated from 1:1980 to 12:2009 using 360 expanding estimation windows starting in 2:1949 for estimating individual models and computing their weights.

Table 11.1 Prior choices

	σ^2	ϕ^a	γ	c
RW	$IG(10^{-6}, 10^{-6})$			
AR	$IG(10^{-6}, 10^{-6})$	$N(0, I)$		
LSTAR	$IG(10^{-6}, 10^{-6})$	$N(0, I)$	$N(3, 0.1)^b$	$N(0, 0.1)^b$

[a] The vector ϕ contains the intercept and autoregression coefficients

[b] We performed a sensitivity analysis. Log marginal likelihoods were computed on the whole data set for AR and LSTAR models with different lag lengths and for a RW model. Then, we multiplied by 5 the prior variance of γ and c and carried out the same calculations. The estimates with the more diffuse prior were marginally lower and the ranking between models almost the same. Furthermore, we computed Bayesian information criteria (that neglect the prior) and also obtained about the same ranking. See [7, Section 3.1] for more details

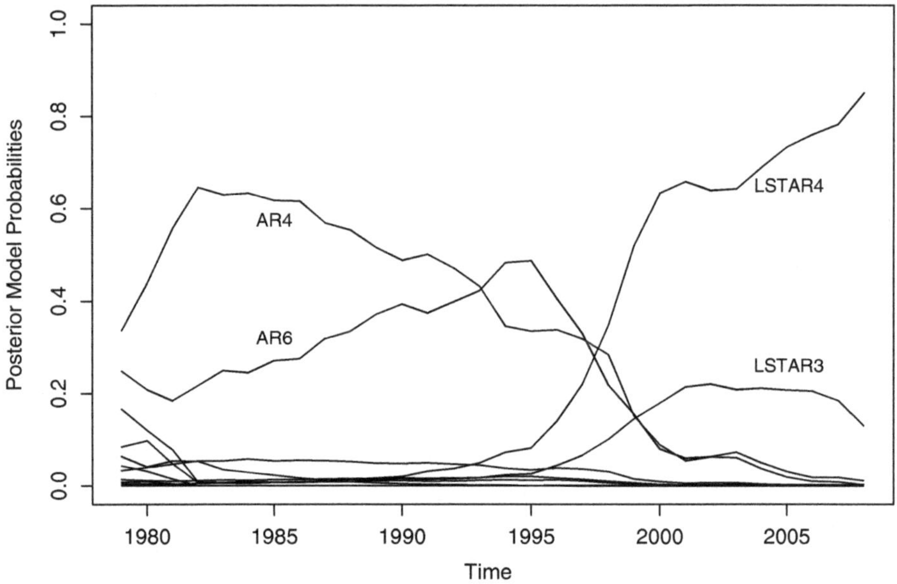

Fig. 11.1 Evolution of PMPs over time for the AR(p) and LSTAR(p), $p = 1, \ldots, 8$. The PMPs are computed once a year over expanding samples starting in 2:1949

11.4 Results

Fig. 11.2 presents the real-time weights produced by the BMA and OP methods over the forecasting period. On the top panel, we see that BMA neglects the RW model and does not select a single model. On the bottom panel, we observe that the OP weights of the AR(4), LSTAR(3), and RW models are almost always equal to zero. Surprisingly, the weights of both methods exhibit a common pattern: linearity is favored until roughly the middle of the forecasting period, while nonlinearity dominates afterward.

We now evaluate predictive performance of our models and model combinations (hereafter our models) with the Diebold–Mariano test, cf. [3], the efficiency test of West and McCracken, cf. [14], and the log score approach. Let the index $t = 1, \ldots, 360$ represent the forecasting period. Table 11.2 shows for our models $\mathrm{MSPE}_k = \frac{1}{360}\sum_{t=1}^{360}(y_t - \bar{y}_{t,k})^2$ and $\mathrm{MAPE}_k = \frac{1}{360}\sum_{t=1}^{360}|y_t - y_{t,k}^{med}|$ where $\bar{y}_{t,k}$ is the predictive mean and $y_{t,k}^{med}$ the predictive median. We see that both criteria give about the same ranking which is dominated by the model averaging methods. The statistical significance of differences between MSPEs or MAPEs is investigated with the Diebold–Mariano test. Tables 11.3 and 11.4 display the robust p-values of this test. We see that BMA outperforms the AR models under both loss structures and that the RW model is beaten by several models under quadratic loss.

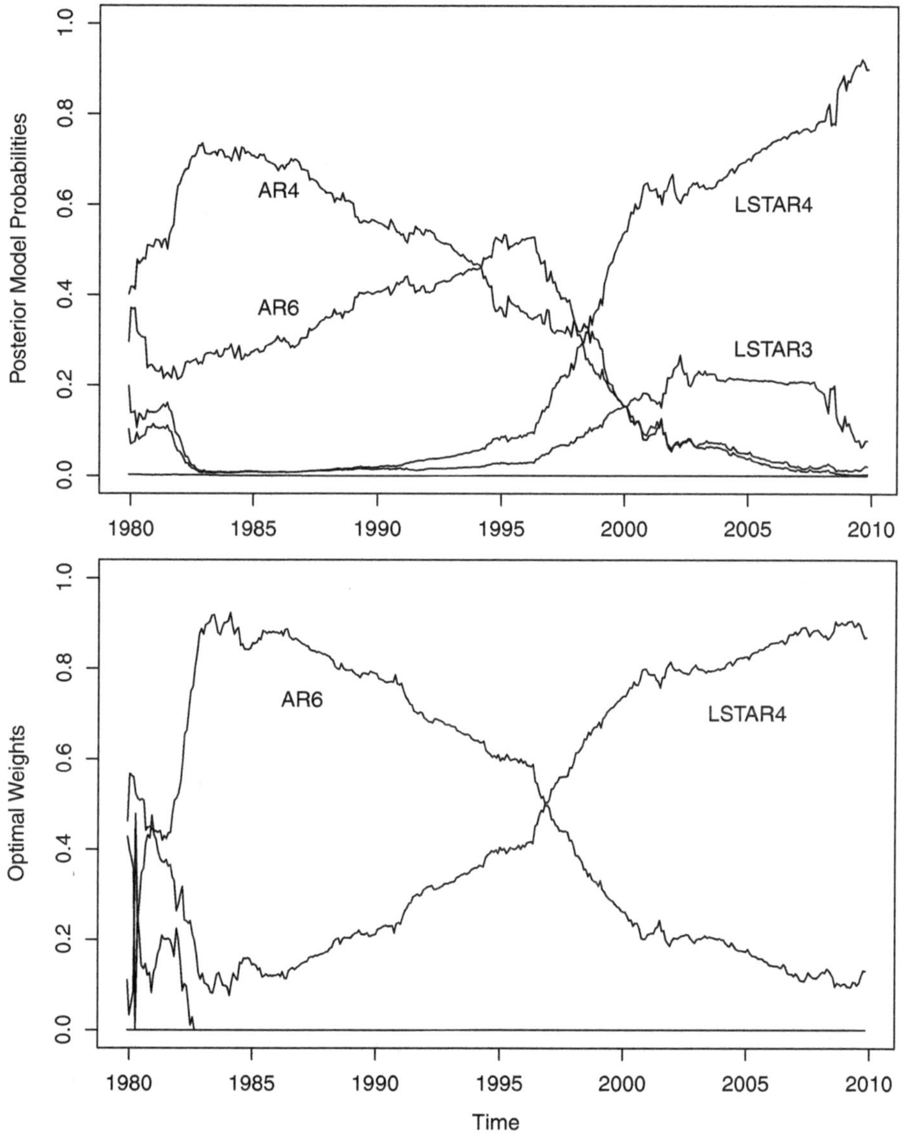

Fig. 11.2 BMA weights (*top panel*) and OP weights (*bottom panel*) allocated to the AR(4), AR(6), LSTAR(3), LSTAR(4), and RW models over the forecasting period

To realize the efficiency test of West and McCracken, we first estimate $y_t = \phi_0 + \phi_1 \tilde{y}_{t,1} + \ldots + \phi_8 \tilde{y}_{t,8} + \varepsilon_t$ where the $\tilde{y}_{t,k}$ are point predictions provided by our models. Then, F-tests of $y_t = \tilde{y}_{t,k} + \varepsilon_t$ against the unrestricted model are performed for all k using a heteroscedasticity and autocorrelation consistent covariance matrix estimate. A model that passes the test is called efficient relative to the others.

Table 11.2 Measures of predictive performance

	MSPE × 100	MAPE
BMA	0.081616	0.022216
OP	0.082526	0.022268
EWMA	0.082587	0.022403
LSTAR(4)	0.083266	0.022513
LSTAR(3)	0.084826	0.022725
AR(4)	0.084995	0.022730
AR(6)	0.085021	0.022778
RW	0.095979	0.022747

Table 11.3 Diebold–Mariano test p-values when using quadratic loss

	OP	EWMA	LSTAR(4)	LSTAR(3)	AR(4)	AR(6)	RW
BMA	0.3435	0.4638	0.5900	0.3214	**0.0428**	**0.0219**	**0.0427**
OP	–	0.9428	0.7399	0.3326	0.2588	0.1029	**0.0646**
EWMA		–	0.7249	0.2732	0.2216	0.1634	**0.0546**
LSTAR(4)			–	0.1665	0.6481	0.6113	0.1153
LSTAR(3)				–	0.9647	0.9565	0.1445
AR(4)					–	0.9858	**0.0586**
AR(6)						–	0.1028

Table 11.4 Diebold–Mariano test p-values when using linear loss

	OP	EWMA	LSTAR(4)	LSTAR(3)	AR(4)	AR(6)	RW
BMA	0.7281	0.3604	0.5009	0.2746	**0.0543**	**0.0349**	0.4961
OP	–	0.3192	0.4365	0.1816	0.1613	**0.0654**	0.5396
EWMA		–	0.7043	0.2724	0.2920	0.1789	0.6431
LSTAR(4)			–	0.3316	0.6951	0.5973	0.7921
LSTAR(3)				–	0.9930	0.9176	0.9790
AR(4)					–	0.8157	0.9792
AR(6)						–	0.9646

Table 11.5 reports the robust p-values of this test performed using predictive means and medians. In both cases, only the BMA and OP methods pass the test at about the 1 % significance level.

Finally, we compute $\text{LS}_{360,k} = \sum_{t=1}^{360} \ln p(y_t|y^{t-1}, M_k)$, where y^{t-1} contains data up to $t-1$ for our different models. The outcomes are displayed in Table 11.6. The LSTAR(4) model obtained the highest log score and the rest of the ranking is again dominated by the model averaging methods. Furthermore, we also present in Fig. 11.3 the evolution of $\text{LS}_{t_1,k} - \text{LS}_{t_1,l}$ for $t_1 = 1,\ldots,360$ for some relevant model pairs. It is noteworthy that the AR(6) model outperforms OP and the LSTAR(4) model at the beginning of the forecasting period and that the cumulative evidence starts to favor the LSTAR(4) over BMA, OP and the AR(6) only in the late 1990s.

Table 11.5 Efficiency test *p*-values

	Using means	Using medians
BMA	0.016072	0.010171
OP	0.012757	0.008844
EWMA	0.004977	0.002310
LSTAR(4)	0.001683	0.001124
LSTAR(3)	0.000102	0.000080
AR(6)	0.000099	0.000046
AR(4)	0.000066	0.000031
RW	0.000000	0.000000

Table 11.6 Predictive performance

	LS_{360}
LSTAR(4)	713.2529
OP	711.8163
BMA	711.0672
EWMA	710.1152
LSTAR(3)	710.1128
AR(6)	708.4054
AR(4)	706.0151
RW	684.1699

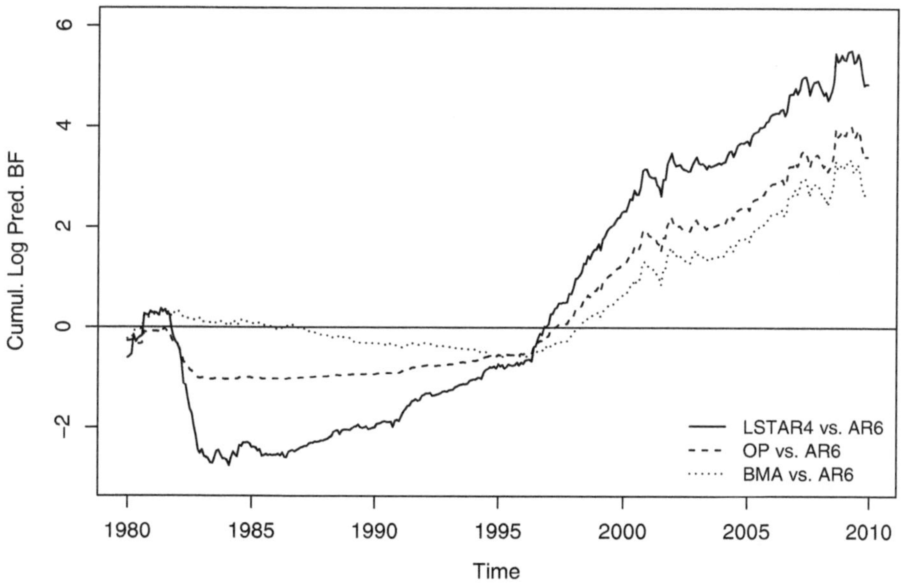

Fig. 11.3 Cumulative log predictive Bayes factors over the forecasting period

11.5 Conclusion

Our initial presumption that linear and nonlinear models approximate the US unemployment DGP in a complementary manner is supported by the dynamic behavior of the BMA and OP weights, by the good predictive performance of the BMA and OP methods and by the cumulative log predictive Bayes factors. It remains difficult to discriminate between the BMA and OP methods. However, only BMA provides a formal treatment of model uncertainty.

Acknowledgements This research relies on Part II of the Ph.D. thesis of the author [7] and is supported by the Department of Management at Fribourg University (Switzerland). The author is very grateful to two anonymous referees for their suggestions, to Mike West, Alessia Paccagnini, and other participants at BAYSM 2014 for their comments and to Philippe J. Deschamps at Fribourg University and David Ardia at Laval University (Canada) for their guidance. He also thanks the Finance, Insurance and Real Estate Department at Laval University for its hospitality. The views expressed in this chapter are those of the author only. The latter is responsible for any remaining errors or shortcomings.

References

[1] Clements, M.P., Smith, J.: Evaluating the forecast densities of linear and nonlinear models: applications to output growth and unemployment. J. Forecast **19**(4), 255–276 (2000)
[2] Deschamps, P.J.: Comparing smooth transition and Markov switching autoregressive models of US unemployment. J. Appl. Econom. **23**(4), 435–462 (2008)
[3] Diebold, F.X., Mariano, R.S.: Comparing predictive accuracy. J. Bus. Econ. Stat. **13**(3), 253–263 (1995)
[4] van Dijk, D., Teräsvirta, T., Franses, P.H.: Smooth transition autoregressive models: a survey of recent developments. Econom. Rev. **21**(1), 1–47 (2002)
[5] Geweke, J., Amisano, G.: Optimal prediction pools. J. Econom. **164**(1), 130–141 (2011)
[6] Geweke, J., Amisano, G.: Prediction with misspecified models. Am. Econ. Rev. **102**(3), 482–486 (2012)
[7] Kolly, J.: Predicting the US unemployment rate using Bayesian model averaging. Ph.D. thesis, Fribourg University, Switzerland (2014). http://doc.rero.ch/record/232699
[8] Koop, G., Potter, S.M.: Dynamic asymmetries in US unemployment. J. Bus. Econ. Stat. **17**(3), 298–312 (1999)
[9] Leamer, E.E.: Specification searches: Ad Hoc inference with nonexperimental data. Wiley, New York (1978)
[10] Meng, X.L., Wong, W.H.: Simulating ratios of normalizing constants via a simple identity: a theoretical exploration. Stat. Sinica **6**(4), 831–860 (1996)
[11] Montgomery, A.L., Zarnowitz, V., Tsay, R.S., Tiao, G.C.: Forecasting the US unemployment rate. J. Am. Stat. Assoc. **93**(442), 478–493 (1998)
[12] Rothman, P.: Forecasting asymmetric unemployment rates. Rev. Econ. Stat. **80**(1), 164–168 (1998)
[13] Teräsvirta, T.: Specification, estimation, and evaluation of smooth transition autoregressive models. J. Am. Stat. Assoc. **89**(425), 208–218 (1994)
[14] West, K.D., McCracken, M.W.: Regression-based tests of predictive ability. Int. Econ. Rev. **39**(4), 817–840 (1998)

Chapter 12
Bayesian Estimation of the Aortic Stiffness based on Non-invasive Computed Tomography Images

Ettore Lanzarone, Ferdinando Auricchio, Michele Conti, and Anna Ferrara

Abstract Aortic diseases are one relevant cause of death in Western countries. They involve significant alterations of the aortic wall tissue, with consequent changes in the *stiffness*, i.e., the capability of the vessel to vary its section secondary to blood pressure variations. In this paper, we propose a Bayesian approach to estimate the aortic stiffness and its spatial variation, exploiting patient-specific geometrical data non-invasively derived from computed tomography angiography (CTA) images. The proposed method is tested considering a real clinical case, and outcomes show good estimates and the ability to detect local stiffness variations. The final objective is to support the adoption of imaging techniques such as the CTA as a standard tool for large-scale screening and early diagnosis of aortic diseases.

Key words: Ordinary differential equations, Parameter estimation, Aortic stiffness, Descending aorta, Computed tomography angiography

12.1 Introduction

Arterial *stiffness*, i.e., the capability of the vessel to vary its section secondary to blood pressure variations, is recognized as a significant predictor of cardiovascular morbidity and mortality [1, 11]. Stiffening of the arterial wall leads to increased systolic and pulse pressures, which in their turn may induce left ventricular hypertrophy and failure, atherosclerosis, as well as aneurysm formation and rupture.

E. Lanzarone (✉)
Istituto di Matematica Applicata e Tecnologie Informatiche (IMATI), Consiglio Nazionale delle Ricerche (CNR), Milan, Italy
e-mail: ettore.lanzarone@cnr.it

F. Auricchio • M. Conti • A. Ferrara
Dipartimento di Ingegneria Civile e Architettura (DICAr), Università degli Studi di Pavia, Pavia, Italy
e-mail: auricchio@unipv.it; michele.conti@unipv.it; anna.ferrara@unipv.it

S. Frühwirth-Schnatter et al. (eds.), *Bayesian Statistics from Methods to Models and Applications*, Springer Proceedings in Mathematics & Statistics 126,
DOI 10.1007/978-3-319-16238-6_12

Hence, the knowledge of the arterial stiffness appears to be fundamental for clinical purposes, but its determination should be performed in a non-invasive manner for large-scale screening and early diagnosis of cardiovascular pathologies.

For a non-invasive estimation, a useful approach consists of an indirect evaluation of the stiffness through *in vivo* measurements of both blood pressure and vessel inner radius over the cardiac cycle. However, at present, this indirect evaluation is performed considering only the maximum and minimum values of pressure and radius waveforms over the cardiac cycle, based on a simple ratio between pressure and radius ranges. Such an approach neglects the information contained in the temporal trend of observed variables and does not include any evaluation about the uncertainty related to the estimated stiffness.

Our goal is to improve the estimation procedure. This work proposes a stochastic method to assess the stiffness of a given aortic region and its spatial variation, exploiting the entire radius and pressure waveforms over the cardiac cycle, and providing the credibility interval associated with the estimates.

Aortic pressure and radius observations are linked through a (linear elastic) *constitutive equation* of the arterial wall, where a constitutive equation is a relation between two physical quantities that is specific to a material and describes the response of the material to external forces. Then, the aortic stiffness is estimated by means of a Bayesian estimation approach able to include the uncertainty of both input variables (pressure and radius) as well as of the arterial stiffness [3, 8]. This methodology has already given good results in other fields, e.g., biology [2], heat transfer [7], and also biomechanics [6]. In this paper, we exploit its potentialities in the arterial stiffness estimation.

12.2 Stiffness Estimation

We consider n cross-sections of an aortic segment. Each section i (with $i = 1, \ldots, n$) is assumed to be a thin-walled circular tube of isotropic linear elastic material with inner radius r_i, thickness h_i, and Young modulus E_i (i.e., the ratio of the stress along an axis to the strain along that axis in the range in which they are proportional). Its constitutive equation is:

$$\mathrm{d}r_i = \frac{r_i^2(t)}{E_i h_i - P_i(t) r_i(t)} \mathrm{d}P_i, \tag{12.1}$$

where $r_i(t)$ and $P_i(t)$ are the state variables observed at section i over time t, whereas $E_i h_i$ is unknown. The latter is assumed as a random quantity given by the sum of a constant expected value $[E_i h_i]_0$ and a Gaussian white noise $\xi_i^E(t)$ scaled by η:

$$E_i h_i = [E_i h_i]_0 + \eta \xi_i^E(t). \tag{12.2}$$

Time t is discretized into instants t_j, and state variables into values $r_{i,j} = r_i(t_j)$ and $P_{i,j} = P_i(t_j)$, respectively. Then, the discretized constitutive equation is solved for $P_{i,j}$ and two further white noises are introduced, related to pressure ($\xi^P_{i,j}$) and radius ($\xi^r_{i,j}$) measurement errors. These noises are assumed additive and proportional to $P_{i,j-1}$ and to the mean between $r_{i,j}$ and $r_{i,j-1}$, respectively.

Hence, Eq. (12.1) is rewritten as:

$$\begin{aligned} P_{i,j} = &\frac{P_{i,j-1} r_{i,j}}{2r_{i,j} - r_{i,j-1}} + \frac{[E_i h_i]_0}{r_{i,j}} \left(1 - \frac{r_i}{2r_{i,j} - r_{i,j-1}}\right) \\ &+ \frac{\eta}{r_{i,j}} \left(1 - \frac{r_i}{2r_{i,j} - r_{i,j-1}}\right) \xi^E_{i,j} + \varepsilon P_{i,j-1} \xi^P_{i,j} + \psi \frac{r_{i,j} + r_{i,j-1}}{2} \xi^r_{i,j} \end{aligned} \quad (12.3)$$

In this way, the conditioned density $f\left(P_{i,j} | P_{i,j-1}, r_{i,j}, r_{i,j-1}, [E_i h_i]_0, \eta^2, \varepsilon^2, \psi^2\right)$ is Gaussian, with:

$$\mu_{i,j} = \frac{P_{i,j-1} r_{i,j}}{2r_{i,j} - r_{i,j-1}} + \frac{[E_i h_i]_0}{r_{i,j}} \left(1 - \frac{r_i}{2r_{i,j} - r_{i,j-1}}\right) \quad (12.4)$$

$$\sigma^2_{i,j} = \frac{\eta^2}{r^2_{i,j}} \left(1 - \frac{r_i}{2r_{i,j} - r_{i,j-1}}\right)^2 + \varepsilon^2 P^2_{i,j-1} + \psi^2 \frac{\left(r_{i,j} + r_{i,j-1}\right)^2}{4} \quad (12.5)$$

Finally, given $m+1$ observations at instants $\{t_0, \ldots, t_j, \ldots, t_m\}$ over the cardiac cycle, the likelihood function is:

$$\begin{aligned} f &\left(\hat{P}_i | \hat{r}_i, [E_i h_i]_0, \eta^2, \varepsilon^2 \psi^2\right) = \\ &= \prod_{j=1}^{m} f\left(P_{i,j} | P_{i,j-1}, r_{i,j}, r_{i,j-1}, [E_i h_i]_0, \eta^2, \varepsilon^2, \psi^2\right) \end{aligned} \quad (12.6)$$

where $\hat{P}_i$ and $\hat{r}_i$ denote the respective set of observations.

Parameters to estimate are $[E_i h_i]_0$ $\forall i$, η^2, ε^2 and ψ^2. Following the Bayesian setting, prior densities are defined for each mentioned parameter.

We assume *a priori* independence among all error parameters (η^2, ε^2, and ψ^2), and between each parameter $[E_i h_i]_0$ and the measurement errors (ε^2 and ψ^2). Moreover, we assume that all parameters $[E_i h_i]_0$ are conditionally independent given η^2.

Then, the choice of the prior densities follows the configuration usually adopted in the literature (e.g., in [6]):

$$\begin{aligned} g\left(\eta^2\right) &= IG\left(\alpha_\eta, \beta_\eta\right) \\ g\left(\varepsilon^2\right) &= IG\left(\alpha_\varepsilon, \beta_\varepsilon\right) \\ g\left(\psi^2\right) &= IG\left(\alpha_\psi, \beta_\psi\right) \\ g\left([E_i h_i]_0 | \eta^2\right) &= N\left([Eh]_0^{prior}, 2\eta^2\right) \end{aligned}$$

Parameters for these prior densities are assigned based on physiological values derived from the literature. Indeed, $[Eh]_0^{prior}$ is set equal to 800 $Pa \cdot m$ considering a Young modulus of 0.4 MPa and a wall thickness of 2 mm. As for the errors, parameters are assigned such that the expected value of $g\left(\eta^2\right)$ is $[Eh]_0^{prior}/10$, the expected value of $g\left(\varepsilon^2\right)$ is 10^4, and the expected value of $g\left(\psi^2\right)$ is 10^{-3} Pa^2/m^2. Hence, $\alpha_\eta = 0.125$, $\beta_\eta = 0.1$, $\alpha_\varepsilon = 0.01$, $\beta_\varepsilon = 0.01$, $\alpha_\psi = 100$, and $\beta_\psi = 10$ are assumed.

12.3 Radius and Pressure Dataset

The evaluation of arterial stiffness according to the proposed stochastic method involves *in vivo* measurements of inner radius and blood pressure waveforms over the cardiac cycle at n cross-sections of the considered arterial district. In our study, we consider eight sections of the aortic arch and the early part of the descending aorta.

In the following, we outline the methodology to derive the dataset from *in vivo* measurements. As discussed below (Sect. 12.3.2), the requirement to keep the entire approach non-invasive leads to include a differential model for generating pressure waveforms.

12.3.1 Radius

Radii $r_{i,j}$ are obtained from patient-specific computed tomography angiography (CTA) images. Recent advances in CTA technology have made four-dimensional (4D) imaging of arterial districts possible, allowing coupling spatial three-dimensional (3D) and temporal information. Indeed, each CTA image is analyzed to get the internal vessel radius of each considered cross-section. Then, the presence of a certain number of images (20 in our case) allows assessing the temporal evolution of radii over the cardiac cycle.

Briefly, the adopted imaging analysis consists of the following three steps:

1. Acquisition of patient-specific medical images.
2. Segmentation and anatomical reconstruction of the 3D lumen profile, using the open source software ITK-Snap (http://www.itksnap.org), which is based on a 3D active contour segmentation method [13]. The segmented models are then exported to stereo-lithography representation (STL format) for the subsequent virtual slicing.
3. Virtual slicing of the 3D reconstruction to get the mean radius at each slice. The slicing procedure is performed as follows:
    - definition of the aortic centerline;
    - definition of n cutting planes normal to the centerline and equally spaced along the centerline;

- detection of the cross-sectional contour points in each plane and spline interpolation;
- calculation of the center of mass for each cross-sectional contour, and computation of the mean radius as the mean value of distances between the center of mass and the contour points.

The entire slicing procedure is implemented through a Python-script exploiting and combining modules of VTK library (http://www.vtk.org) and of VMTK library (http://www.vmtk.org).

12.3.2 Pressure

Direct non-invasive measures of blood pressure in central arteries, e.g., in the aorta, are not feasible nowadays; in fact, direct measurement requires catheterization, which is usually performed only during surgery and not in the clinical routine.

Alternatively, two indirect approaches can be followed for obtaining pressure in central arteries: to generate the pressure waveforms by means of an appropriate mathematical model of the arterial circulation, or to derive the aortic pressure from peripheral measurements. However, also in the second case, the central aortic pressure is derived from the peripheral one by means of a mathematical model.

We follow the first alternative, and blood pressures $P_{i,j}$ are generated using a lumped parameter model of the arterial circulation, based on [4, 5].

Such a model describes the arterial tree from the aortic valve to the capillaries by means of 63 large artery segments and 30 peripheral networks, which are appropriately connected in parallel and series to reproduce structure of the arterial system. Each segment is represented by an electrical circuit, in which tension and electric current are the analogous of pressure and flow in the vessel, respectively. Thus, each segment is characterized by two ordinary differential equations, one for blood pressure and one for blood flow. Moreover, each segment is characterized by resistances, inductances, and compliances, whose values are given by the vessel geometrical and mechanical properties [12]. The input of the overall tree is the blood flow waveform through the aortic valve.

Numerically solving the equations, the temporal evolution of pressure in each segment is obtained. Then, pressure waveforms corresponding to the considered cross-sections are taken, and values at the instants of CTA images are extracted.

12.4 Application to a Real Clinical Case and Results

The proposed approach is applied to a real clinical case, considering an elderly female patient with a descending aorta dilation, probably related to an aneurysm, which suggests a localized vessel stiffening. Figures 12.1 and 12.2 schematically

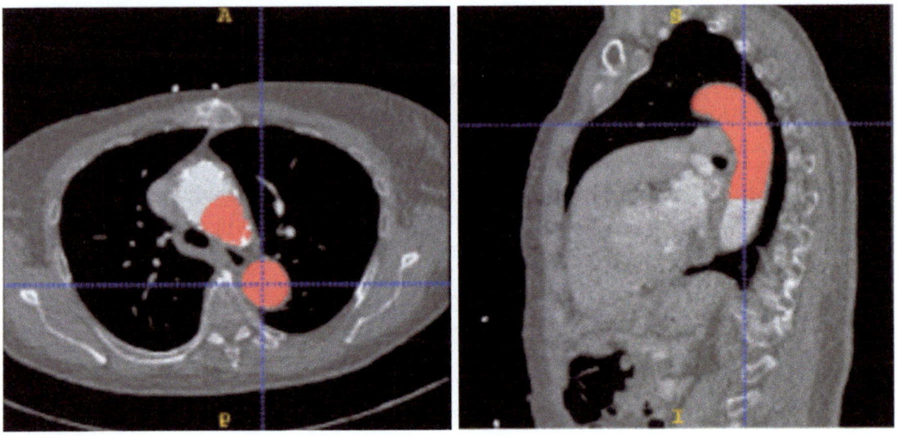

Fig. 12.1 Segmentation of an acquired CTA image using open source software ITK-Snap

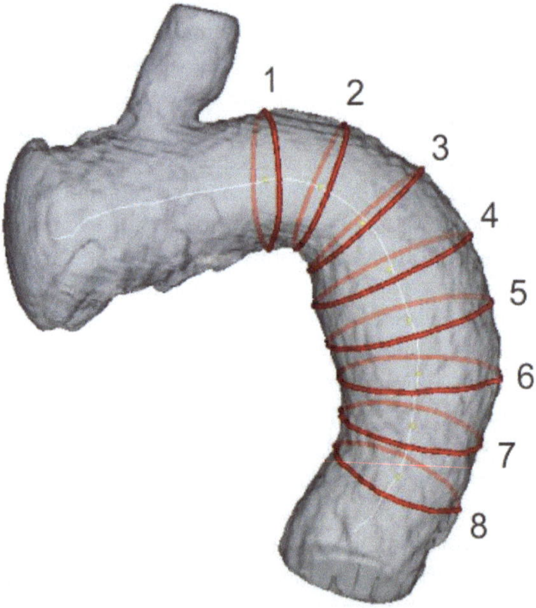

Fig. 12.2 3D reconstruction of the considered aorta lumen using open source software ITK-Snap. The district is virtually sliced in eight equally spaced sections starting from the left subclavian artery. The aortic centerline used for slicing is also represented

show the segmentation of the aorta and the 3D reconstruction of the aorta lumen, for one of the acquired CTA images. Then, the obtained radii at the eight cross-sections are reported in Table 12.1.

As for the lumped parameters model, the peripheral resistances are increased by 40 % with respect to the original values [4, 5] in order to consider the observed patient's hypertension. The obtained pressures at the eight cross-sections are reported in Table 12.2.

Table 12.1 Cross-sectional radii [mm] at the eight cross-sections

Time %	Sect. 1	Sect. 2	Sect. 3	Sect. 4	Sect. 5	Sect. 6	Sect. 7	Sect. 8
0	12.91	12.77	13.95	15.37	15.43	14.41	12.87	12.46
5	13.16	12.97	14.10	15.55	15.68	14.43	12.95	12.61
10	13.31	13.18	14.54	15.76	15.89	14.54	13.20	12.76
15	13.36	13.35	14.57	16.02	15.97	14.67	13.18	12.91
20	13.38	13.39	14.60	16.12	16.11	14.78	13.32	13.00
25	13.47	13.48	14.71	16.13	16.12	14.78	13.36	12.94
30	13.49	13.42	14.67	16.05	16.12	14.71	13.27	12.89
35	13.56	13.48	14.57	16.00	16.01	14.67	13.24	12.84
40	13.47	13.42	14.38	15.86	15.95	14.64	13.21	12.78
45	13.35	13.29	14.37	15.79	15.88	14.57	13.22	12.76
50	13.30	13.20	14.35	15.73	15.74	14.51	13.11	12.74
55	13.18	13.07	14.12	15.57	15.70	14.50	13.05	12.69
60	13.14	13.01	14.05	15.49	15.71	14.43	13.04	12.60
65	13.11	12.95	14.03	15.47	15.60	14.41	13.00	12.56
70	13.07	12.88	13.93	15.35	15.56	14.38	12.96	12.57
75	13.01	12.86	13.90	15.29	15.47	14.35	12.93	12.54
80	12.96	12.80	13.85	15.27	15.41	14.32	12.88	12.49
85	12.95	12.75	13.77	15.16	15.40	14.21	12.85	12.39
90	12.84	12.66	13.70	15.11	15.36	14.24	12.86	12.40
95	12.83	12.72	13.71	15.14	15.34	14.26	12.78	12.39
100	12.91	12.77	13.95	15.37	15.43	14.41	12.87	12.46

Time is expressed in percentage with respect to the cardiac cycle (equal to 0.8 s), and the first and the last observations coincide due to the periodic cycle

Estimates are obtained by means of a Gibbs sampling scheme, implemented in JAGS [10] with 200,000 iterations, a burn-in of 10,000 iterations, and a thinning interval of 10. Satisfactory traceplots are obtained, thus indicating the convergence of the Markov chain.

Estimates of the products $[E_i h_i]_0$ for all i are obtained, and Young moduli E_i are then derived assuming $h_i = 2\ mm\ \forall i$. The spatial trend of E_i along with the considered aortic segment is plotted in Fig. 12.3, in terms of posterior mean and posterior standard deviation. Estimates consistent with the literature [9] and with other deterministic techniques are found. Moreover, the stiffness spatial variation is caught, in agreement with the characteristics of the considered clinical case where a localized stiffening was expected at some sections. Finally, other prior hyperparameters around the adopted ones have been tested, and the posterior estimates are not affected by them.

Table 12.2 Cross-sectional pressures [mmHg] at the eight cross-sections

Time %	Sect. 1	Sect. 2	Sect. 3	Sect. 4	Sect. 5	Sect. 6	Sect. 7	Sect. 8
0	92.34	92.31	92.27	92.24	92.20	92.17	92.13	92.10
5	101.98	101.08	100.20	99.26	98.37	97.56	96.83	96.14
10	121.46	121.22	120.95	120.61	120.24	119.84	119.41	118.87
15	133.55	133.44	133.31	133.14	132.95	132.75	132.53	132.27
20	138.92	139.08	139.23	139.39	139.54	139.68	139.81	139.94
25	141.30	141.59	141.87	142.17	142.47	142.76	143.05	143.35
30	140.69	141.04	141.38	141.75	142.11	142.46	142.80	143.17
35	135.81	136.24	136.66	137.12	137.57	138.00	138.41	138.84
40	133.26	133.19	133.13	133.08	133.04	133.03	133.04	133.08
45	132.46	132.56	132.65	132.75	132.85	132.93	133.01	133.08
50	129.72	129.78	129.83	129.89	129.95	130.00	130.06	130.11
55	126.20	126.20	126.19	126.19	126.19	126.19	126.18	126.18
60	122.77	122.68	122.59	122.48	122.37	122.26	122.15	122.03
65	119.09	118.94	118.79	118.61	118.44	118.27	118.09	117.90
70	114.77	114.71	114.65	114.57	114.49	114.40	114.32	114.21
75	109.84	109.81	109.79	109.76	109.74	109.71	109.69	109.67
80	105.12	105.19	105.26	105.34	105.42	105.51	105.59	105.68
85	101.15	101.25	101.36	101.48	101.61	101.73	101.86	102.00
90	97.71	97.80	97.89	98.00	98.10	98.21	98.31	98.43
95	94.71	94.74	94.76	94.79	94.82	94.86	94.89	94.94
100	92.34	92.31	92.27	92.24	92.20	92.17	92.13	92.10

Time is expressed in percentage with respect to the cardiac cycle (equal to 0.8 s), and the first and the last observations coincide due to the periodic cycle

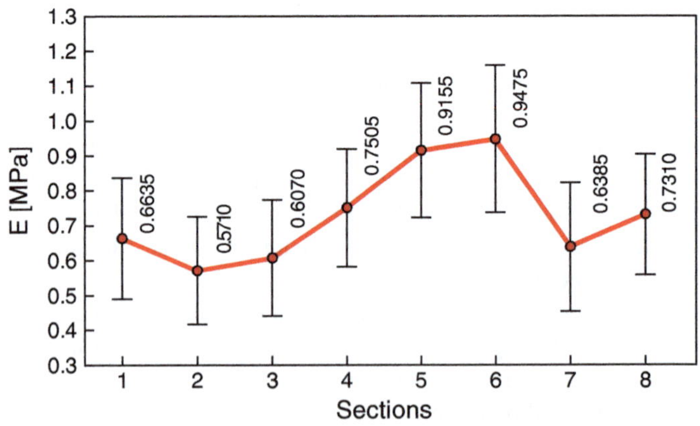

Fig. 12.3 Young modules E_i estimated at each section i with an assumed wall thickness $h_i = 2$ mm $\forall i$: posterior means and error bars equal to posterior standard deviations

12.5 Conclusions

In this paper, we propose a non-invasive approach to estimate the aortic stiffness of a specific subject/patient and its regional changes. The aortic stiffness is computed combining cross-sectional radial enlargements of the aorta with the respective pressure waveforms, taking into account their entire waveforms over the cardiac cycle. In particular, cross-sectional radii are obtained elaborating 4D CTA images, whereas pressures are simulated using a lumped parameter model of the arterial circulation to keep the methodology non-invasive. Finally, the approach exploits a Bayesian estimation method to include the uncertainty of both the input variables, i.e., vessel radii and blood pressures, and the estimated stiffness.

Results are promising, and computational times for obtaining estimates once CTA images are stored (including the time for dataset generation) are limited to some seconds. Low computational times of the proposed methodology are fundamental for large-scale application, thus ensuring a practical clinical application of the method.

Future work will be conducted for considering more complex constitutive equations able to better detail the 3D structure of the vessels, whereas the equation adopted in this paper refers to a section of a cylindrical incompressible vessel. This means that, when applied to cases in which a sudden variation of radius is present, the movement of largest section could be slowed by the neighboring ones and result in a stiffness overestimation. Moreover, we will investigate the possibility of coupling the Bayesian estimation with more complex computational analyses, e.g., the Finite Element Analysis.

Acknowledgements The authors acknowledge Flagship Project "Factory of the Future Fab@Hospital", funded by Italian CNR and MIUR organizations. Michele Conti acknowledges ERC Starting Grant through the Project ISOBIO: Isogeometric Methods for Biomechanics (No. 259229).

References

[1] Dernellis, J., Panaretou, M.: Aortic stiffness is an independent predictor of progression to hypertension in non-hypertensive subjects. Hypertension **45**, 426–431 (2005)

[2] Gilioli, G., Pasquali, S., Ruggeri, F.: Bayesian analysis of a stochastic predator-prey model with nonlinear functional response. Math. Biosci. Eng. **9**, 75–96 (2012)

[3] Kloeden, P.E., Platen, E.: Numerical Solution of Stochastic Differential Equations. Springer, Berlin (1992)

[4] Lanzarone, E., Liani, P., Baselli, G., Costantino, M.L.: Model of arterial tree and peripheral control for the study of physiological and assisted circulation. Med. Eng. Phys. **29**, 542–555 (2007)

[5] Lanzarone, E., Casagrande, G., Fumero, R., Costantino, M.L.: Integrated model of endothelial NO regulation and systemic circulation for the comparison between pulsatile and continuous perfusion. IEEE Trans. Bio-Med. Eng. **56**, 1331–1340 (2009)

[6] Lanzarone, E., Ruggeri, F.: Inertance estimation in a lumped-parameter hydraulic simulator of human circulation. J. Biomech. Eng. Trans. ASME **135**, 061012 (2013)

[7] Lanzarone, E., Pasquali, S., Mussi, V., Ruggeri, F.: Bayesian estimation of thermal conductivity and temperature profile in a homogeneous mass. Numer. Heat Transfer B Fund. **66**, 397–421 (2014)

[8] Oksendal, B.: Stochastic Differential Equations: An Introduction with Applications, 6th edn. Springer, Berlin (2003)

[9] Pearson, A.C., Guo, R., Orsinelli, D.A., Binkley, P.F., Pasierski, T.J.: Transesophageal echocardiographic assessment of the effects of age, gender, and hypertension on thoracic aortic wall size, thickness, and stiffness. Am. Heart J. **128**, 344–351 (1994)

[10] Plummer, M.: JAGS: a program for analysis of Bayesian graphical models using Gibbs sampling. In: Proceedings of the 3rd International Workshop on Distributed Statistical Computing, Vienna, Austria (2003)

[11] Quinn, U., Tomlinson, L.A., Cockcroft, J.R.: Arterial stiffness. J. Roy. Soc. Med. **1**, 1–18 (2012)

[12] Westerhof, N., Bosman, F., Vries, C.J.D., Noordergraaf, A.: Analog studies of the human systemic arterial tree. J. Biomech. **56**, 121–143 (1969)

[13] Yushkevich, P.A., Piven, J., Hazlett, H., Smith, R., Ho, J.G.S., Gerig, G.: User-guided 3d active contour segmentation of anatomical structures: significantly improved efficiency and reliability. Neuroimage **31**, 1116–1128 (2006)

Chapter 13
Bayesian Filtering for Thermal Conductivity Estimation Given Temperature Observations

Laura Martín-Fernández and Ettore Lanzarone

Abstract International standards often require complex experimental layouts to estimate the thermal conductivity of materials, and they marginally take into account the uncertainty in the estimation procedure. In this paper, we propose a particle filtering approach coupled with a simple experimental layout for the real-time estimation of the thermal conductivity in homogeneous materials. Indeed, based on the heat equation, we define a state-space model for the temperature evaluation based on the unknown conductivity, and we apply a Rao-Blackwellized particle filter. Finally, the approach is validated considering heating and cooling cycles given to a specimen made up of polymethylmethacrylate (PMMA) in forced convection. Results show good estimates in accordance with the PMMA conductivity range, and computational times confirm the possibility of a real-time estimation.

Key words: Thermal conductivity estimation, Temperature transient method, Real-time estimation, Rao-Blackwellized particle filter

13.1 Introduction

The thermal conductivity of materials is an important parameter for handling and choosing materials in contexts where the thermal behaviour needs to be considered or controlled. This occurs in different fields (e.g., thermal insulation, environmental heating or cooling, heat exchange) and in several industrial applications. Unfortunately, the thermal conductivity of materials is not always known *a priori*, and needs to be estimated when a specific specimen of a material is being considered. Thus, a cheap conductivity estimation procedure would be a useful tool for several

L. Martín-Fernández (✉)
Departamento de Física Aplicada, Universidad de Granada, Granada, Spain
e-mail: lauramartin@ugr.es

E. Lanzarone
Istituto di Matematica Applicata e Tecnologie Informatiche (IMATI), Consiglio Nazionale delle Ricerche (CNR), Milan, Italy
e-mail: ettore.lanzarone@cnr.it

S. Frühwirth-Schnatter et al. (eds.), *Bayesian Statistics from Methods to Models and Applications*, Springer Proceedings in Mathematics & Statistics 126,
DOI 10.1007/978-3-319-16238-6_13

applications. Moreover, the quantification of the uncertainty associated with the estimates would allow a robust analysis of thermal behaviour, based on stochastic material parameters.

International standards propose several methods for estimating the thermal conductivity of materials [2, 5, 6]. However, they require complex experimental layouts and marginally consider uncertainty in the estimation procedure.

To improve the approach, a Bayesian estimation of conductivity coupled with a simple experimental layout has recently been proposed in [8]. This exploits MCMC simulation for obtaining the conductivity posterior density, possibly including the generation of latent temperatures in points where temperature is not acquired. However, this approach requires acquiring all temperature measurements before data are processed and estimates are provided.

In this paper, we propose a Rao-Blackwellized Particle Filter (PF) that allows real-time estimation instead of waiting for the entire dataset. The algorithm has already been applied to parameter estimation in ordinary differential equations [9, 10]. In this work, we apply it to the heat equation, i.e., a partial differential equation. The aim is to exploit the benefits of the simple experimental layout already proposed while integrating real-time estimation.

13.2 Method

In this section, we first describe the state-space model we refer to, which is derived from the unidirectional heat equation. Then, we present the application of the Rao-Blackwellized PF to this state-space model.

13.2.1 State-Space Model

We consider the following nonlinear state-space model, which is derived from the spatial and temporal discretizations of the unidirectional heat equation, as in [8].

$$\mathbf{T}_j = \mathbf{T}_{j-1} + \mathbf{g}_j \lambda_0 + \mathbf{Q}_j \Delta \mathbf{w}_j, \tag{13.1}$$

$$\mathbf{o}_j = \mathbf{T}_j + \mathbf{H}_j \boldsymbol{\xi}_j, \tag{13.2}$$

where:

- $\mathbf{T}_j$ is the vector of true temperatures at each discretized point at time j ($j = 1, \ldots, F$);
- λ_0 is the thermal conductivity to estimate;
- $\mathbf{g}_j = \tau \mathbf{L}_{j-1}$;
- τ is the time interval of the temporal discretization;
- $\mathbf{Q}_j = \eta \mathbf{D}_{\mathbf{L}_{j-1}}$;

- $\Delta\mathbf{w}_j$ is a vector of independent Wiener processes;
- $\mathbf{o}_j$ is the corresponding vector of noisy temperature observations;
- $\mathbf{H}_j = \varepsilon\mathbf{D}_{\mathbf{T}_j}$;
- η and ε are the errors related to the noise processes;
- $\boldsymbol{\xi}_j$ is a vector of independent white noise processes;
- $\mathbf{D_X}$ denotes a diagonal matrix with $d_{i,i} = x_i$;
- $\mathbf{L}_{j-1} = \mathbf{A}\mathbf{T}_{j-1} + \mathbf{b}_{j-1}$.

Moreover,

$$\mathbf{A} = -\frac{1}{\rho c h^2}\begin{bmatrix} 2 & -1 & 0 & \dots & \dots & \dots & 0 \\ -1 & 2 & -1 & 0 & \dots & \dots & 0 \\ 0 & -1 & 2 & -1 & 0 & \dots & 0 \\ \vdots & \ddots & \ddots & \ddots & \ddots & \ddots & \vdots \\ 0 & \dots & 0 & -1 & 2 & -1 & 0 \\ 0 & \dots & \dots & 0 & -1 & 2 & -1 \\ 0 & \dots & \dots & \dots & 0 & -1 & 2 \end{bmatrix}_{N\times N}, \quad \mathbf{b}_j = \frac{1}{\rho c h^2}\begin{bmatrix} T_{0,j} \\ 0 \\ \vdots \\ 0 \\ T_{N+1,j} \end{bmatrix}_{N\times 1},$$

where N is the number of discretized internal points, $T_{0,j}$ and $T_{N+1,j}$ are the temperatures at the surface points, h is the distance between two consecutive points in the material, ρ and c are the material density and specific heat, respectively.

13.2.2 Rao-Blackwellized Particle Filter

The prior of $\mathbf{T}_0$ is defined as a delta measure that is centred at the corresponding acquired value and, based on Eq. (13.1), it is expressed by

$$p\left(\mathbf{T}_j|\mathbf{T}_{j-1},\lambda_0\right) = N\left(\mathbf{T}_{j-1} + \mathbf{g}_j\lambda_0, \tau\mathbf{Q}_j\mathbf{Q}_j^{\top}\right). \tag{13.3}$$

We remark that, as the temperature dynamics depend on the unknown parameter λ_0, the process $\mathbf{T}_j$ is not Markovian.

Considering $\Delta\mathbf{T}_j = \mathbf{T}_j - \mathbf{T}_{j-1}$, Eq. (13.1) can be rewritten as a linear-Gaussian state-space model with state variable λ_0 and observation vector $\Delta\mathbf{T}_j$:

$$\begin{aligned} \lambda_{0,j} &= \lambda_{0,j-1}, \\ \Delta\mathbf{T}_j &= \mathbf{g}_j\lambda_{0,j} + \mathbf{Q}_j\Delta\mathbf{w}_j. \end{aligned} \tag{13.4}$$

The conditional density of $\Delta\mathbf{T}_j$ given $\lambda_{0,j}$ is expressed by a Gaussian density:

$$p\left(\Delta\mathbf{T}_j|\lambda_{0,j}\right) = N\left(\mathbf{g}_j\lambda_{0,j}, \tau\mathbf{Q}_j\mathbf{Q}_j^{\top}\right).$$

Moreover, we assume that λ_0 is *a priori* Gaussian, with mean value $\hat{\lambda}_{0,0}$ and variance P_0. Then, its posterior density at time j is also Gaussian:

$$p\left(\lambda_0|\Delta\mathbf{T}_{1:j}\right) = N\left(\hat{\lambda}_{0,j}, P_j\right),$$

with

$$\hat{\lambda}_{0,j} = \int \lambda_0 \, p\left(\lambda_0|\Delta\mathbf{T}_{1:j}\right) \, d\lambda_0$$

and

$$P_j = \int \left(\lambda_0 - \hat{\lambda}_{0,j}\right)^2 p\left(\lambda_0|\Delta\mathbf{T}_{1:j}\right) \, d\lambda_0.$$

Due to the state-space model of Eq. (13.4) being linear in λ_0 and the Gaussian likelihood of $\lambda_{0,j}$, we can apply a Kalman filter [1, 7] to exactly compute the posterior distribution of λ_0.

In this paper, we apply a Rao-Blackwellized PF (RBPF) [3, 4] to jointly approximate the posterior distribution of the temperatures and estimate the unknown conductivity λ_0. The proposed PF handles S particles; hence, a set of S Kalman filters running in parallel is implemented.

In this way,

$$p\left(\mathbf{T}_j|\mathbf{T}_{0:j-1}^{(i)}, \mathbf{o}_{1:j-1}\right) = N\left(\boldsymbol{\beta}_j^{(i)}, \mathbf{B}_j^{(i)}\right), \tag{13.5}$$

where:

$$\boldsymbol{\beta}_j^{(i)} = \mathbf{g}_j^{(i)} \hat{\lambda}_{0,j-1}^{(i)} + \mathbf{T}_{j-1}^{(i)},$$

$$\mathbf{B}_j^{(i)} = \mathbf{g}_j^{(i)} P_{j-1}^{(i)} \mathbf{g}_j^{(i)\top} + \tau \mathbf{Q}_j^{(i)} \mathbf{Q}_j^{(i)\top},$$

and superscript $^{(i)}$ indicates the particle $\mathbf{T}_{0:j}^{(i)}$.

Finally, the posterior estimate of λ_0 at each time j is obtained using the statistics generated by the RBPF. In particular, the posterior mean and variance of λ_0 conditional on the observations $\mathbf{o}_{1:j}$ can be approximated as

$$\hat{\lambda}_{0,j}^S = \sum_{i=1}^{S} v_j^{(i)} \hat{\lambda}_{0,j}^{(i)},$$

$$P_j^S = \sum_{i=1}^{S} v_j^{(i)} \left[\left(\hat{\lambda}_{0,j}^{(i)} - \hat{\lambda}_{0,j}^S\right)^2 + P_j^{(i)}\right],$$

where $v_j^{(i)}$ is the importance weight related to the particle $\mathbf{T}_j^{(i)}$.

In this way, a real-time estimation that evolves over time is achieved.

13.3 Application to the Experimental Layout

The proposed approach has been applied to estimate the thermal conductivity of a polymer, for which a conductivity range of variability is known due to the polymer family, whereas the specific value within this range is unknown.

The experimental layout, as in [8], consists of a specimen made up of PMMA, with square faces (side of 20 cm) and thickness of 15 cm. Seven thermocouples are put in the specimen in the centre of the square faces along a line: two external thermocouples on the square faces and five equally spaced internal thermocouples within the specimen. Finally, lateral rectangular faces are thermally insulated to guarantee unidirectional heat flow. A picture of the specimen is reported in Fig. 13.1.

Three experiments have been conducted (A, B and C): each time a heating and cooling cycle, lasting about 40 h, has been given to the specimen in forced convection, within a range where the PMMA conductivity is constant (see [6], Part 5). Temperature signals were acquired from the thermocouples with a frequency of 10 Hz, and then digitalized. A time interval of 60 s (1 min) was considered; thus, the temperature observations were taken directly every 600 digitalized values, without any moving average or filtering. As example, temperature trends acquired in experiment B are reported in Fig. 13.2.

Fig. 13.1 PMMA specimen with square faces, seven inserted thermocouples, and lateral rectangular faces thermally insulated

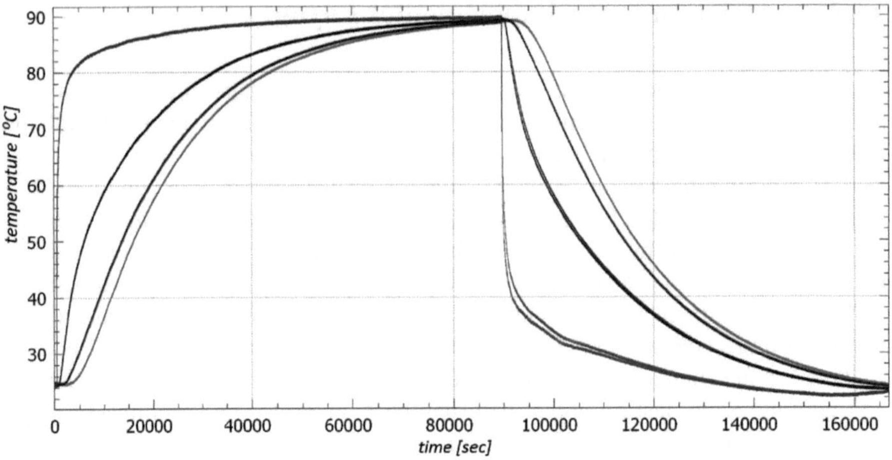

Fig. 13.2 Acquired trends in experiment B: faster dynamics refer to boundary temperatures, whereas slower dynamics are observed while moving towards the centre of the specimen

The proposed RBPF has been implemented in Visual Basic using the Microsoft Net Framework 4. Then, the computations have been run on a Microsoft Windows machine with 8 cores and 15 GB RAM, which was installed on a server with an AMD Opteron 6328 processor. We have applied the algorithm with $S = 600{,}000$ particles, $\tau = 60\,\text{s}$, $\eta = 0.00015$, and $\varepsilon = 0.0015$.

We have first validated the approach considering several simulated datasets of 2 h, in which boundary and initial conditions are taken from the experiments, whereas internal temperatures are simulated with different conductivities within the PMMA range.

Then, we have applied the proposed RBPF to the real datasets of the experiments, considering the entire experiment duration.

13.4 Results

As for the validation phase, results show estimation errors always lower than 1 %. Indeed, Table 13.1 shows the estimated conductivity for three different simulated values of conductivity λ_0^*, and for both a period of 2 h during the heating phase and a period of 2 h during the cooling phase.

Considering the application to the real acquired datasets, results in Table 13.2 show stable estimations among the experiments within the PMMA range, associated with low variances.

As for experiment A, we also show the plots of the λ_0 estimation over time (Fig. 13.3a) and the estimated internal temperatures compared to the observed ones (Fig. 13.3b–f). Results show that the estimated value of λ_0 is stabilized after few time instants, thus allowing low errors in temperature estimates already from the first points.

Table 13.1 Posterior mean and variance of λ_0 estimates (in $\frac{W}{mK}$) obtained by the proposed PF algorithm for the six simulated datasets considered

		$\lambda_0^* = 0.18$	$\lambda_0^* = 0.20$	$\lambda_0^* = 0.22$
Heating	λ_0 mean $\left[\frac{W}{mK}\right]$	0.1808	0.2002	0.2204
	λ_0 variance $\left[\frac{W^2}{m^2K^2}\right]$	$5.16 \cdot 10^{-12}$	$2.04 \cdot 10^{-9}$	$6.05 \cdot 10^{-12}$
Cooling	λ_0 mean $\left[\frac{W}{mK}\right]$	0.1791	0.2013	0.2197
	λ_0 variance $\left[\frac{W^2}{m^2K^2}\right]$	$7.84 \cdot 10^{-9}$	$8.49 \cdot 10^{-12}$	$2.42 \cdot 10^{-8}$

Table 13.2 Posterior mean and variance of λ_0 estimates (in $\frac{W}{mK}$) obtained by the proposed PF algorithm for the real datasets of the three experiments

	Experiment A	Experiment B	Experiment C
λ_0 mean $\left[\frac{W}{mK}\right]$	0.2044	0.2097	0.2169
λ_0 variance $\left[\frac{W^2}{m^2K^2}\right]$	$2.64 \cdot 10^{-14}$	$3.14 \cdot 10^{-14}$	$3.21 \cdot 10^{-14}$

13.5 Discussion and Conclusions

In this paper, we have applied a PF for the joint estimation of thermal conductivity and temperatures in a solid homogeneous material. The linear structure of the model and the Gaussian noise processes allowed us to apply a Rao-Blackwellized PF that uses a bank of Kalman filters for the analytical integration of the unknown parameter, when approaching the posterior density of the temperatures.

The proposed approach seems to be able to improve the estimation procedure of thermal conductivity in homogeneous materials, by equipping the layout of [8] with a real-time estimation.

The estimated distribution of conductivity and the temperature trajectories show satisfactory fits to the data, thus confirming the goodness of the proposed method. Moreover, the thermal conductivity value is stabilized after few time instants, allowing to accurately follow the temperature profile along the experiment.

Real-time estimation is also confirmed, since the computational time to process a time step is lower than the time step itself.

Future work will be conducted to include the joint estimation of latent temperatures in points where a thermocouple is not inserted, to generalize the particle-filter approach to non-homogeneous materials (by considering a specific thermal conductivity for each acquired point that may vary over time), and to evaluate the behaviour when applied to two-dimensional and three-dimensional problems.

Acknowledgements This work has been supported by the "Consejería de Economía, Innovación, Ciencia y Empleo de la Junta de Andalucía" of Spain under Project TIC-03269.

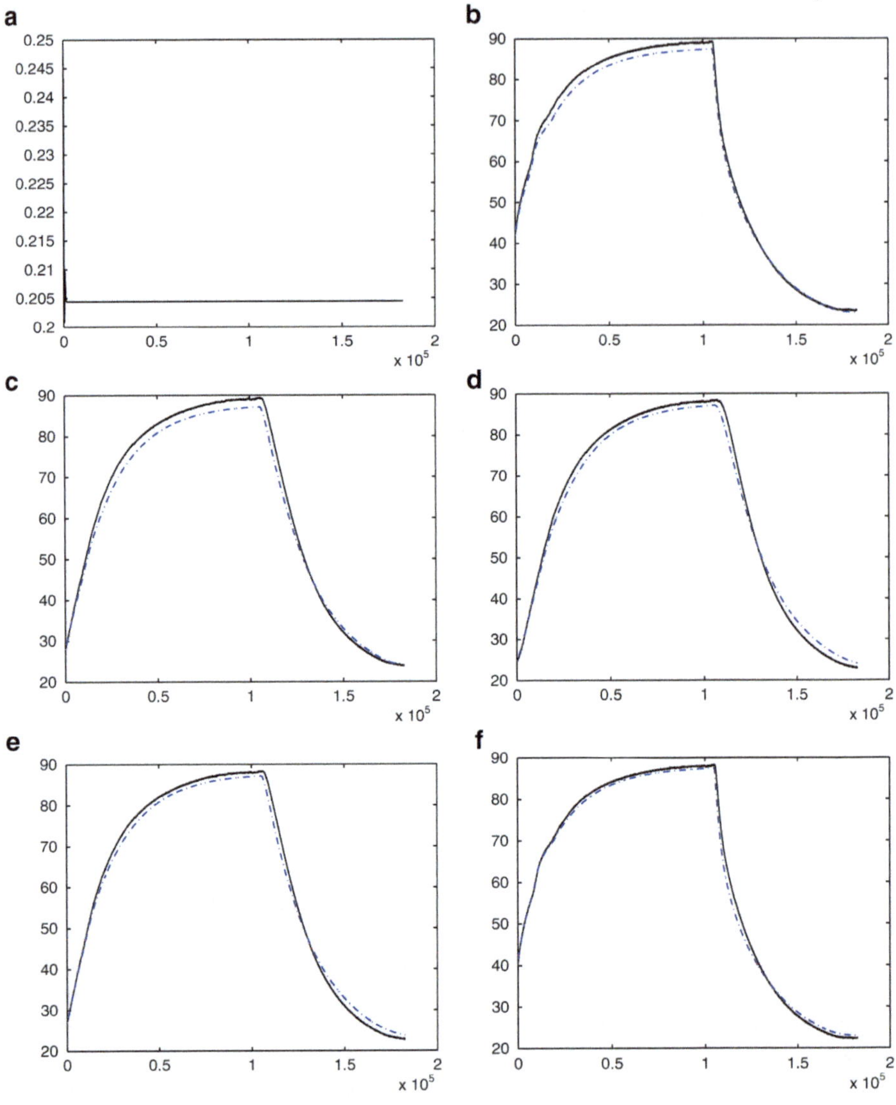

Fig. 13.3 (**a**) mean $\lambda_{0,j}$ (in $\frac{W}{mK}$) over time instants j (in seconds); (**b**)–(**f**) internal estimated (*dotted lines*) and acquired (*solid lines*) temperatures (in °C) over time instants j (in seconds)

References

[1] Andrieu, C., Doucet, A., Holenstein, R.: Particle Markov chain Monte Carlo methods. J. Roy. Stat. Soc. B Meth. **72**, 269–342 (2010)

[2] ASTM E1952 - 11. Standard test method for thermal conductivity and thermal diffusivity by modulated temperature differential scanning calorimetry. http://www.astm.org

[3] Chen, R., Liu, J.S.: Mixture Kalman filters. J. Roy. Stat. Soc. B Meth. **62**, 493–508 (2000)
[4] Doucet, A., Godsill, S., Andrieu, C.: On sequential Monte Carlo sampling methods for Bayesian filtering. Stat. Comput. **10**, 197–208 (2000)
[5] EN 12667. Thermal performance of building materials and products. Determination of thermal resistance by means of guarded hot plate and heat flow meter methods. Products of high and medium thermal resistance. http://www.en-standard.eu
[6] ISO 22007-1:2009. Plastics - determination of thermal conductivity and thermal diffusivity. http://www.iso.org
[7] Kalman, R.E.: A new approach to linear filtering and prediction problems. J. Basic Eng. **82**, 35–45 (1960)
[8] Lanzarone, E., Pasquali, S., Mussi, V., Ruggeri, F.: Bayesian estimation of thermal conductivity and temperature profile in a homogeneous mass. Numer. Heat Transfer B Fund. **66**, 397–421 (2014)
[9] Martín-Fernández, L., Gilioli, G., Lanzarone, E., Míguez, J., Pasquali, S., Ruggeri, F., Ruiz, D.P.: Joint parameter estimation and biomass tracking in a stochastic predator-prey system. In: Lanzarone, E., Ieva, F. (eds.) Springer Proceedings in Mathematics & Statistics - The Contribution of Young Researchers to Bayesian Statistics, vol. 63, pp. 23–27. Springer International Publishing, Switzerland (2014)
[10] Martín-Fernández, L., Gilioli, G., Lanzarone, E., Míguez, J., Pasquali, S., Ruggeri, F., Ruiz, D.P.: A Rao-Blackwellized particle filter for joint parameter estimation and biomass tracking in a stochastic predator-prey system. Math. Biosci. Eng. **11**, 573–97 (2014)

Chapter 14
A Mixture Model for Filtering Firms' Profit Rates

Ellis Scharfenaker and Gregor Semieniuk

Abstract Existing methods for sample selection from noisy profit rate data in the industrial organization field of economics tend to be conditional on a covariate's value that risks discarding valuable information. We condition sample selection on the profit rate data structure instead by use of a Bayesian mixture model. In a two-component (signal and noise) mixture that reflects the prior belief of noisy data, each firm profit rate observation is assigned an indicator latent variable. Gibbs sampling determines the latent variables' posterior densities, sorting profit rate observations to the signal or noise component. We apply two model specifications to empirical profit rate cross-sections, one with a normal and one with a Laplace signal component. We find the Laplace specification to have a superior fit based on the Bayes factor and the profit rate sample to be time stationary Laplace distributed, corroborating earlier estimates of cross-section distributions. Our model retains 97 %, as opposed to as little as 20 %, of the raw data in a previous application.

Key words: Mixture model, Sample selection, Laplace distribution, Profit rates, Gibbs sampler

14.1 Introduction

The formation of a general rate of profit, around which profit rates of competitive firms gravitate, was stressed by classical political economists beginning with Adam Smith [11], who theorized that through competition and capital mobility a tendency of the equalization of profit rates across all competitive industries would emerge. At the same time, individual firms would be able to increase their profit rates through innovations and cost-cutting, until new entrants into their market would

E. Scharfenaker (✉) • G. Semieniuk
Department of Economics, The New School for Social Research, New York, NY, USA
e-mail: schae854@newschool.edu; semig068@newschool.edu

S. Frühwirth-Schnatter et al. (eds.), *Bayesian Statistics from Methods to Models and Applications*, Springer Proceedings in Mathematics & Statistics 126,
DOI 10.1007/978-3-319-16238-6_14

compete the profit rate back down. This would lead to a cross-section of profit rates where most of the firms would have profit rates close to the average. Neoclassical economic theory assumes equal profit rates and therefore a degenerate distribution under perfect competition [12].

The shape of *empirical* cross-sectional distributions of profit rates has been scarcely investigated; however, [2] finds Laplace profit rate cross-sections in a small sample of only long-lived firms from *Thomson Datastream* data. One problem with the study of profit rates is that the data is "noisy." An observed profit rate of more than 50 % would raise the eyebrows of most economists, let alone the profit rates greater than 1,000 % or smaller than −1,000 % present in the data. In order to effectively rid the dataset of observations far from the mode, [2] discard all firms that live less than 27 years, the time period spanned by their unbalanced panel. With this method they retain only 20 % of their original data. Sample selection based on a covariate such as age or size for studying the distributions of firm characteristics is justified under the belief that small or young firms belong to an entirely different set of data that are subject to separate "entry and exit" constraints [2]. Other studies start from a preselected data set of only large firms [1, 3, 8]. However, the essentially arbitrary determination of what is "long lived" or "large" may prevent an understanding of how the large majority of competitive firm profit rates are distributed, since a large share of firms are small or short-lived. The implicit prior when applying this method is that young or small firms may produce noisy outliers that arise from possible accounting problems when studying ratios of variables such as the profit rate. We believe a more flexible method for sample selection that explicitly models the noise and signal can improve this line of research.

We construct a Bayesian mixture model to select a sample based on the profit data structure rather than a cut-off in a covariate. To do this, we continue to assume that cross-sections of profit rate observations are distributed as a mixture of a peaked signal that contains most of the data and a flat noise component, but it is not known which observations belong to which mixture component. This can be found by assigning an indicator latent variable to each firm and estimating the marginal posterior densities of each of them: a high mean posterior density assigns the observation to the signal distribution, a low one to noise. We find closed form full conditional posterior densities for all but one parameter and estimate the remaining parameter's posterior density numerically. Applying Gibbs sampling from the set of conditional posterior densities, we make posterior inference about each observation being in the signal or noise component. By selecting a plausible mixture model informed by economic theory and previous research, we find that Bayesian methods far outperform the existing non-statistical selection methods in terms of data conservation. Further, our results have significant implications for future work in the distributional analysis of economic variables as our method of data selection is easily extendable.

14.2 Model Specification

As discussed in Sect. 14.1, economic theory suggests profit rate cross-sections are unimodally distributed with only moderate mass in the tails and previous empirical research has highlighted the problems of outliers. A mixture model with two components, one peaked, one flat, incorporates this prior knowledge. The flat noise component models the outliers and the peaked signal component models the profit rates of firms from which to select the sample.

14.2.1 Observation Model and Priors

Two model specifications $M_{\mathscr{N}}$ and $M_{\mathscr{L}}$ are considered. The signal density, f_s, for model $M_{\mathscr{N}}$ is normal, for model $M_{\mathscr{L}}$ Laplace. The noise density, f_n is (diffusely) normal. Then for either model a weighted mixture density f of the profit rate distribution with parameter blocks θ_s and θ_n and N data points $Y = (y_1, \ldots, y_n)$ is

$$f(y_i|q, \theta_s, \theta_n) = q\, f_s(y_i|\theta_s) + (1-q)\, f_n(y_i|\theta_n), \tag{14.1}$$

with q being a weighing parameter. Assigning a latent variable, $Z = (z_1, \ldots, z_n)$ with $z_i \in \{0, 1\}\, \forall\, i$ to each observation gives the equivalent specification

$$f(y_i|q, z_i, \theta_s, \theta_n) = f_s(y|\theta_s)^{z_i}\, f_n(y_i|\theta_n)^{(1-z_i)}, \tag{14.2}$$

where $z_i \sim Bernoulli(q)$. If $z_i = 1$, then observation i is in the signal component, if $z_i = 0$, it is noise. For the two models the specifications are

$$f_{\mathscr{N}}(y_i|q, \mu_{\mathscr{N}}, \sigma_{\mathscr{N}}, \mu_n, \sigma_n) = qN(y_i|\mu_{\mathscr{N}}, \sigma_{\mathscr{N}}) + (1-q)N(y_i|\mu_n, \sigma_n), \tag{14.3}$$

$$f_{\mathscr{L}}(y_i|q, \mu_{\mathscr{L}}, \sigma_{\mathscr{L}}, \mu_n, \sigma_n) = qL(y_i|\mu_{\mathscr{L}}, \sigma_{\mathscr{L}}) + (1-q)N(y_i|\mu_n, \sigma_n). \tag{14.4}$$

To sample from the full marginal posterior densities of all parameters we choose appropriate priors. For most parameters, these are conjugate priors:

$$\pi(q) \sim U(0,1), \quad \pi(\mu_{\mathscr{N}}|\chi, \xi) \sim N(\chi, \xi), \quad \pi(\sigma_{\mathscr{N}}|\alpha, \beta) \sim IG(\alpha, \beta), \tag{14.5}$$

$$\pi(\mu_n|\nu, \upsilon) \sim N(\nu, \upsilon), \quad \pi(\sigma_n|\delta, \gamma) \sim IG(\delta, \gamma). \tag{14.6}$$

However, the Laplace distribution with density

$$f(y|\mu, \sigma) = \frac{1}{2\sigma} e^{-|y-\mu|\sigma^{-1}}, \tag{14.7}$$

is not a member of the exponential family [9]. Priors are chosen that admit a closed posterior form for the scaling parameter $\sigma_{\mathscr{L}}$ and the location parameter choice follows [10, p. 20] which can be used to represent the posterior density as an $N+1$ component mixture of truncated normal distributions.[1]

$$\pi(\mu_{\mathscr{L}}|\tau) \sim N(0,\tau), \quad \pi(\sigma_{\mathscr{L}}|\phi,\psi) \sim IG(\phi,\psi)\,. \tag{14.8}$$

14.2.2 Inference

We use the Gibbs sampler in the class of Markov chain Monte Carlo methods for posterior inference. Without requiring direct knowledge of the intractable joint posterior density of the parameter vector, $\boldsymbol{\theta}$, the Gibbs sampler produces a sequence of each parameter $\{\theta_j^{(g)}\}_{g=1}^G$ from its full conditional posterior density. This is a Markov chain whose stationary distribution is that of the joint posterior density [4, 7]. For inference with the mixture model it is therefore sufficient to specify all full conditional posterior densities, $\pi(\theta_j|\theta_{k\neq j},Y)$. Denoting the count of latent variables equal one as $\sum_{i=1}^N z_i = M$ and the count of latent variables equal zero as $\sum_{i=1}^N (1-z_i) = N-M = K$, the full conditional posterior densities for weighing and noise parameters are

$$\pi(q|\mu_s,\sigma_s,\mu_n,\sigma_n,Z,Y) \sim Beta(M+1,\ K+1), \tag{14.9}$$

$$\pi(\mu_n|\sigma_n,Z,\nu,\upsilon,Y) \sim N\left(\frac{\nu^{-1}+\sigma_n^{-1}\sum_{i=1}^N(1-z_i)y_i}{\frac{1}{\upsilon}+\frac{K}{\sigma_n}},\ \frac{1}{\frac{1}{\upsilon}+\frac{K}{\sigma_n}}\right), \tag{14.10}$$

$$\pi(\sigma_n|\mu_n,Z,\delta,\gamma,Y) \sim IG\left(\delta+\frac{1}{2}K,\ \gamma+\frac{1}{2}\sum_{i=1}^N(1-z_i)(y_i-\mu_n)^2\right), \tag{14.11}$$

for both models. For the model $M_{\mathscr{N}}$ the full conditional posteriors are

$$\begin{aligned}\pi(z_i=1|q,\mu_{\mathscr{N}},\sigma_{\mathscr{N}},\mu_n,\sigma_n,Z_{j\neq i},Y)&\\ \sim Bern\left(\frac{qN(y_i|\mu_{\mathscr{N}},\sigma_{\mathscr{N}})}{qN(y_i|\mu_{\mathscr{N}},\sigma_{\mathscr{N}})+(1-q)N(y_i|\mu_n,\sigma_n)}\right)&\end{aligned} \tag{14.12}$$

$$\pi(\mu_{\mathscr{N}}|\sigma_{\mathscr{N}},Z,\chi,\xi,Y) \sim N\left(\frac{\chi^{-1}+\sigma_{\mathscr{N}}^{-1}\sum_{i=1}^N z_iy_i}{\frac{1}{\xi}+\frac{M}{\sigma_{\mathscr{N}}}},\ \frac{1}{\frac{1}{\xi}+\frac{M}{\sigma_{\mathscr{N}}}}\right) \tag{14.13}$$

$$\pi(\sigma_{\mathscr{N}}|\mu_{\mathscr{N}},Z,\alpha,\beta,Y) \sim IG\left(\alpha+\frac{1}{2}M,\ \beta+\frac{1}{2}\sum_{i=1}^N(y_i-\mu_{\mathscr{N}})^2\right)\,. \tag{14.14}$$

[1] We thank Christian Robert for pointing out this possibility during the BAYSM'14 conference.

For model $M_{\mathscr{L}}$ the full conditional posteriors are additionally

$$\pi(z_i|q,\mu_{\mathscr{L}},\sigma_{\mathscr{L}},\mu_n,\sigma_n,Z_{j\neq i},Y) \sim Bern\left(\frac{qL(y_i|\mu_{\mathscr{L}},\sigma_{\mathscr{L}})}{qL(y_i|\mu_{\mathscr{L}},\sigma_{\mathscr{L}})+(1-q)N(y_i|\mu_n,\sigma_n)}\right), \tag{14.15}$$

$$\pi(\sigma_{\mathscr{L}}|\mu_{\mathscr{L}},Z,\phi,\psi,Y) \sim IG(\phi+M,\ \psi+\sum_{i=1}^{N} z_i(y_i-\mu_{\mathscr{L}})), \tag{14.16}$$

$$\pi(\mu_{\mathscr{L}}|\sigma_{\mathscr{L}},Z,\tau,Y) \sim c\prod_{i=1}^{N} L(\mu_{\mathscr{L}}|y_i,\sigma_{\mathscr{L}})^{z_i} N(0,\tau). \tag{14.17}$$

Equation (14.17) requires estimation of the posterior over a grid of points, and this approximation is used both for Gibbs sampling and marginal posterior density estimation. All other posterior densities are in closed form.

After G rounds of sampling from the Gibbs sampler, the Gibbs output can be used to make a numerical estimate of the marginal likelihood of each model, $\pi(y|M_i)$. The marginal likelihood of a model with a given vector of parameters θ^* is

$$\pi(y) = \frac{\pi(y|\theta^*)\pi(\theta^*)}{\pi(\theta^*|y)}. \tag{14.18}$$

Chib [5] has shown that the Gibbs output can be used to estimate $\pi(y)$, if full conditional posterior densities are available for each of K parameters by writing posterior densities as

$$\pi(\theta^*|y) = \pi(\theta_1^*|y)\times\pi(\theta_2^*|y,\theta_1^*)\times\ldots\times\pi(\theta_K^*|y,\theta_1^*,\ldots\theta_{K-1}^*), \tag{14.19}$$

which can be estimated from the Gibbs sampler by resampling for each parameter

$$\pi(\theta_1^*|y) = \frac{1}{G}\sum_{g=1}^{G}\pi(\theta_1^*|y,\theta_2^g\ldots\theta_K^{(g)}),$$

$$\pi(\theta_2^*|y,\theta_1^*) = \frac{1}{G}\sum_{g=1}^{G}\pi(\theta_2^*|y,\theta_1^*,\theta_3^{(g)}\ldots\theta_K^{(g)}), \text{ and so forth.} \tag{14.20}$$

We use this method to estimate the marginal likelihood, using the approximation from (14.17) for the conditional posterior density estimate of $\mu_{\mathscr{L}}$. The next section presents results from applying this inference procedure to a new dataset.

14.3 Results

14.3.1 Data

The data is from COMPUSTAT comprising US stock market-listed companies in years 1962–2012, for which profit rate cross-sections have not yet been analyzed.[2] We calculate the profit rate by dividing the difference of net sales and operating costs, which equals operating income before depreciation, by total assets.[3] Government as well as financial services, real estate, and insurance have been excluded because the former does not partake in competition and the latter adheres to different accounting conventions for revenue calculation that makes this part of the industries incomparable. Outliers in some years with absolute value above 10,000 % have also been excluded to make inference comparable between years with outliers three orders of magnitude larger and most other years with all outliers within the bounds. Therefore, our data is comprised of firms under the standard industrial classification (SIC) numbers 1,000–6,000 and 7,000–9,000, a total of 279,891 observations with on average 5,500 annual observations, a 3.5 times larger dataset than previously analyzed profit rate data [2].

14.3.2 Posterior Inference for Two Models

The Gibbs sampler is run 1,000 times for every year for both models, and the first 100 burn-in iterations are discarded. All chains converge quickly to the posterior distribution and diagnostics show stationarity for all years.[4] The boxplots in Fig. 14.1 show narrow posterior distributions for all parameters in every year for both models. The value of q between 90 % and 100 % indicates that more than nine tenths of firm observations are sorted into the signal distribution. The changing location of the parameter values over time is of economic interest but is beyond the scope of the current paper. The parameters from the noise component of the mixture (not shown) settle on very different values over the years depending on the nature of the noise in each year. It is noticeable that model $M_{\mathcal{N}}$ samples lower q parameters on average, indicating that this model assigns a higher fraction of observations to the noise component. This should be expected, given that the Laplace distribution has fatter tails than the normal distribution and that there are large outliers in the data. The question is which model fits the data better.

[2] Although the data is available from 1950, as [6] points out, there is a serious selection bias in pre-1962 data that is tilted toward big historically successful firms.

[3] These are COMPUSTAT items ($SALES$), ($XOPR$), and (AT).

[4] See Appendix for details.

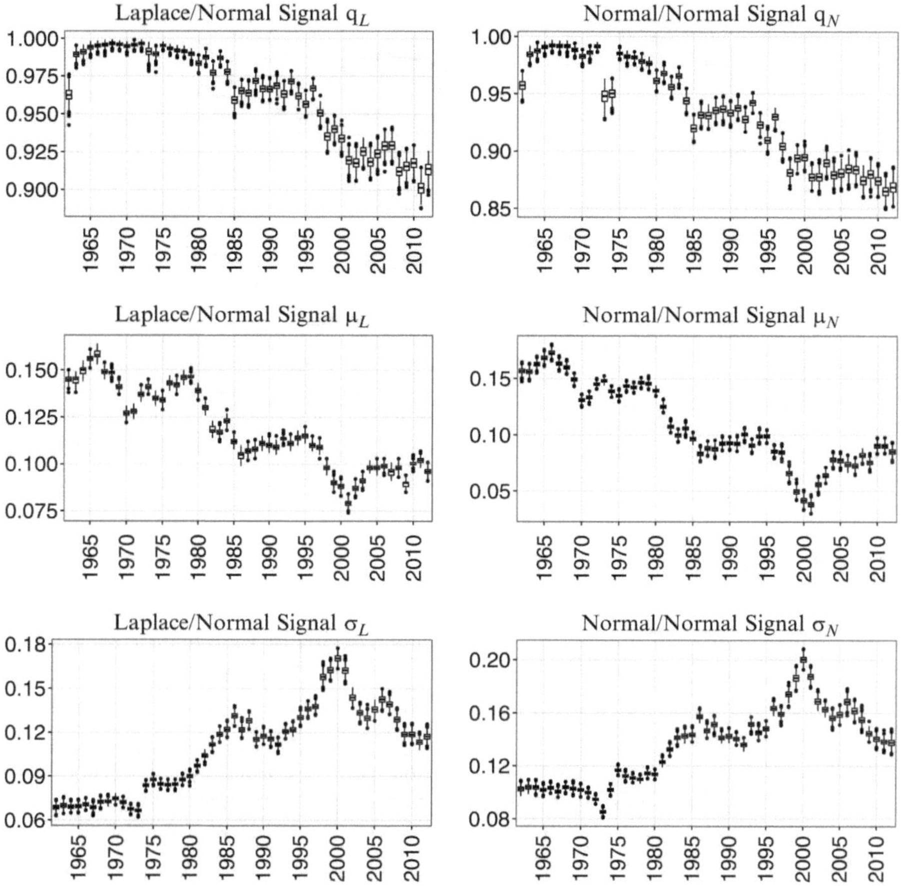

Fig. 14.1 Boxplots of Gibbs output for the signal distribution f_s, with weighing parameter q (*top*), location parameter μ (*center*), and scale parameter σ (*bottom*) for every year for both models. The *left panel* shows results for the Laplace signal model, the *right panel* for the normal signal model

14.3.3 Model Comparison

In order to select a model, Fig. 14.2 compares log marginal likelihoods (LML) of the two models. Results for model $M_{\mathscr{L}}$ are robust to respecifications of the grid for (14.17). The Bayes factor ranges over the orders of magnitude 1,700 and 5,300. The data clearly increase the odds in favor of the Laplace model, $M_{\mathscr{L}}$, in every single year. Therefore, we analyze the signal distribution obtained by the model $M_{\mathscr{L}}$.

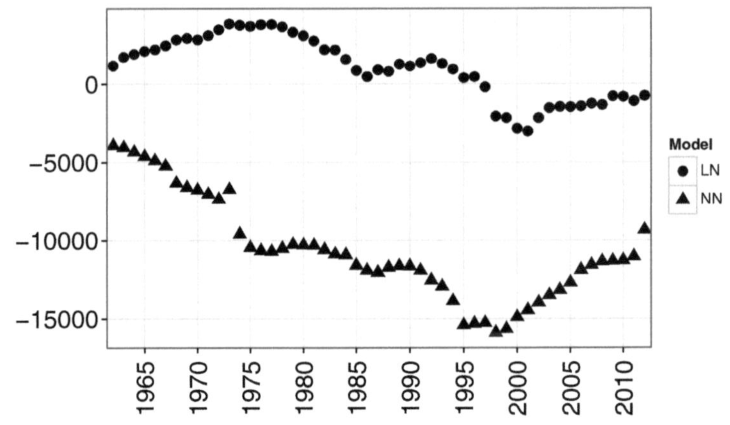

Fig. 14.2 Log marginal likelihoods of Laplace model (*circles*) and normal model (*triangles*). The marginal likelihood favors the Laplace model in every year

14.3.4 The Filtered Distribution

The Gibbs sampler of model $M_{\mathscr{L}}$ also yields a series of samples for each latent variable z_i with realizations of either one or zero. A choice has to be made about what share of ones in each series qualifies the corresponding observation, y_i to be assigned to the signal. The motivation of the model is to keep as much of the original data as possible, so we adopt a simple filter rule $\mathscr{A}$, whereby all observations whose latent variable posterior mean is greater than 0.05 belong to the signal.[5] This ensures discarding only observations that lie far away from the mode. Since [2] has suggested that 20 % of their data are Laplace distributed, we apply a second filter rule $\mathscr{B}$ that keeps only those observations in the signal whose latent variable posterior mean is above 0.99. Only observations close to the Laplace density are retained.

Table 14.1 shows that both choice rules remove all extreme outliers and shrink the minimum and maximum toward the median. Filter rule $\mathscr{A}$ retains 97 % of the data, and the more restrictive filter rule $\mathscr{B}$ retains 90 %. It is also evident that there is no perceptible change in the interquartile range and that the 95th percentile has a similar value for all samples. Only the fifth percentile differs, which is clarified below.

Figure 14.3 shows the signal distribution on a log density scale for a selection of years under the permissive choice rule $\mathscr{A}$ in the top panel, and the restrictive choice rule $\mathscr{B}$ in the bottom panel. Ninety percent of the data are Laplace (the Laplace or double exponential density appears as a tent shape for a density plot

[5] Discarding z_i whose share below any value in the range 0.01–0.9 corresponds to discarding 3–5 % of the data. Therefore, the data filter is not very sensitive to permissive choice rules other than 0.05.

Table 14.1 Summary statistics including several quantiles, of pooled raw profit rates (r) and signal profit rates after discarding all observation with latent posterior mean below 0.05 (rule $\mathscr{A}$) and after discarding all observation with latent posterior mean below 0.99 $\mathscr{B}$

	Min	5 Perc.	1st Qu.	Median	Mean	3rd Qu.	95 Perc.	Max	% Retained
r	−99.50	−0.615	0.027	0.112	−0.047	0.174	0.298	99.49	100
$r_{\mathscr{A}:0.05}$	−1.657	−0.381	0.038	0.114	0.067	0.174	0.296	1.801	97.25
$r_{\mathscr{B}:0.99}$	−1.657	−0.125	0.038	0.114	0.067	0.175	0.292	1.801	90.01

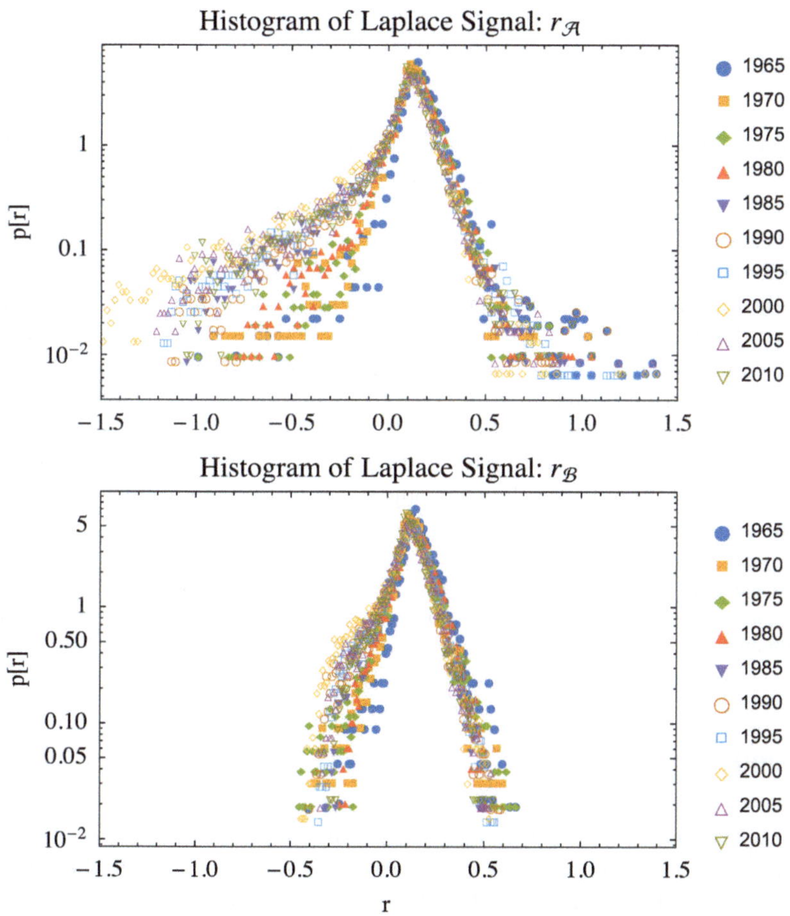

Fig. 14.3 Histogram plots on a log density scale of profit rates (r) for a selection of cross-sections after filtering for permissive (*top*) and restrictive (*bottom*) choice rules. At least 90 % of data are Laplace distributed and another 7 % add a negative skew

with a log scale). The next 7 % of the data add negative skewness, giving rise to the different fifth percentile in the summary statistics table. A much larger share of observations, including younger firms, are Laplace. Only 6 % of the data would remain,if we had applied the previously used decision rule of keeping only firms alive over the 51 years spanned by the data set (or 33 % for firms surviving at least 27 years). This shows that the mixture model retains much more data and therefore information about firm competition than a selection based on a covariate such as firm age.

14.4 Conclusion

We construct a Bayesian mixture model as a sample selection tool that explicitly models the component of microeconomic data typically discarded as noise. In this model, latent variables assign observations to either signal or noise based on the specification of the priors and the data structure. By Gibbs sampling from all full conditional posterior distributions, we find posterior densities for the latent variables that allow us to select a sample of firms without confounding noise for economic analysis. Unlike previous work, the latent variable Gibbs sampler procedure yields a distribution without having to make additional assumptions about minimum size or age. Comparison of marginal likelihoods favored a Laplace signal distribution. We select a sample containing more than 97 % of the data allowing us to investigate the distributional form of profit rates in the U.S. economy arising from competition between firms. This filtering logic is applicable to other univariate data sets that include confounding outliers.

Acknowledgements We would like to thank Duncan Foley, Sid Dalal, Sid Chib, and two anonymous referees for providing helpful guidance for this research. The data was made available by Columbia University Library.

Appendix

To show stationarity of the Markov chains from our Gibbs sampler we display the trace plots for a selection of years in Fig. 14.4. The stationarity of all Markov chains is supported by the Geweke and Heidelberger–Welch diagnostics.

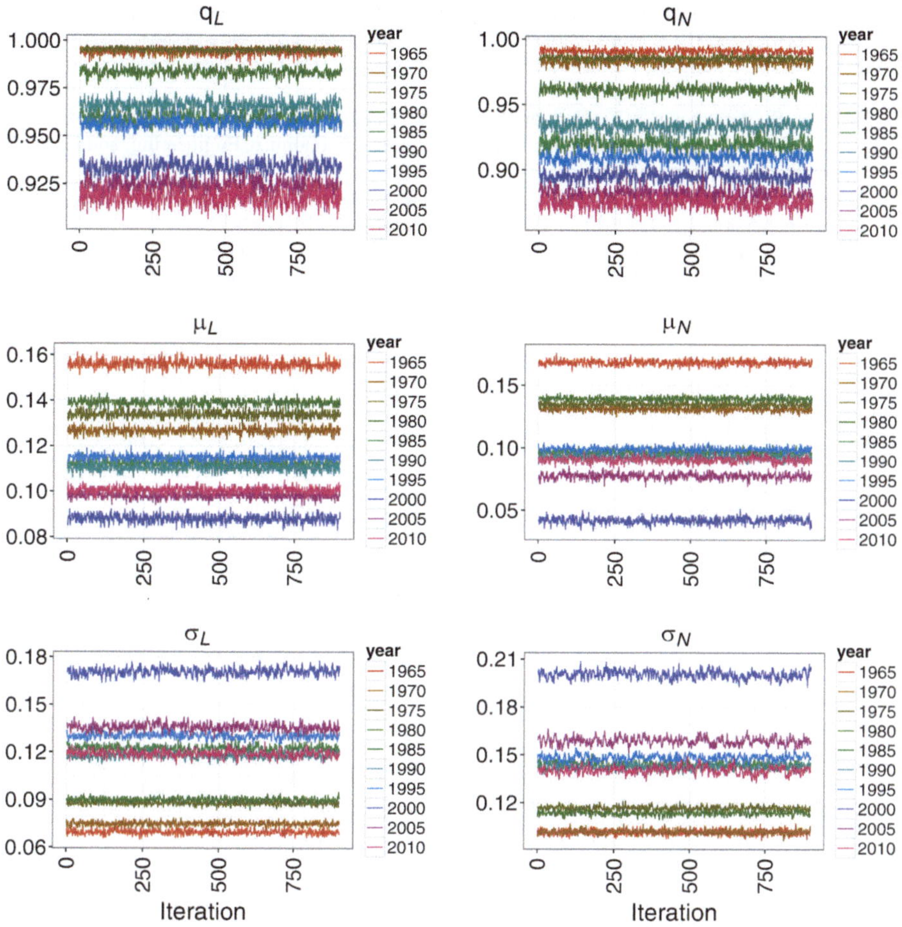

Fig. 14.4 Trace plots from Gibbs sampler for Laplace signal (*left*) and normal signal (*right*) after discarding the initial 100 iterations

References

[1] Alfarano, S., Milaković, M.: Does classical competition explain the statistical features of firm growth? Econ. Lett. **101**(3), 272–274 (2008)

[2] Alfarano, S., Milaković, M., Irle, A., Kauschke, J.: A statistical equilibrium model of competitive firms. J. Econ. Dyn. Control **36**(1), 136–149 (2012)

[3] Bottazzi, G., Dosi, G., Lippi, M., Pammolli, F., Riccaboni, M.: Innovation and corporate growth in the evolution of the drug industry. Int. J. Ind. Organ. **19**(7),1161–1187 (2001)

[4] Casella, G., George, E. I.: Explaining the Gibbs sampler. Am. Stat. **46**(3), 1–11 (1992)

[5] Chib, S.: Marginal likelihood from the Gibbs output. JASA **90**(432), 1313–1321 (1995)

[6] Fama, E.F., French, K.R.: The cross-section of expected stock returns. J. Finance **47**(2), 427–465 (1992)

[7] Geman, S., Geman, D.: Stochastic relaxation, Gibbs distributions, and the Bayesian restoration of images. IEEE Trans. Pattern Anal. Mach. Intell. **6**(6), 721–741 (1984)
[8] Hymer, S., Pashigian, P.: Firm size and rate of growth. J. Political Economy **7**(6), 556–569 (1962)
[9] Kotz, S., Kozubowski, T., Podgórski, K.: The Laplace Distribution and Generalizations: A Revisit with New Applications to Communications, Economics, Engineering, and Finance. Birkhauser, Boston (2001)
[10] Marin, J.M., Pillai, N., Robert, C.P., Rousseau, J.: Relevant statistics for Bayesian model choice, version, 21 Oct 2011. http://www.arxiv.org/abs/1110.4700v1
[11] Smith, A.:The Wealth of Nations. Penguin, London/New York (1982)
[12] Varian, H.: Microeconomic Analysis, 3rd edn. W.W. Norton and Company, New York/London (1992)

Index

S. Frühwirth-Schnatter et al. (eds.), *Bayesian Statistics from Methods to Models and Applications*, Springer Proceedings in Mathematics & Statistics 126,
DOI 10.1007/978-3-319-16238-6

R

S

T

U

W

MIX
Papier aus verantwortungsvollen Quellen
Paper from responsible sources
FSC® C105338

If you have any concerns about our products,
you can contact us on
ProductSafety@springernature.com

In case Publisher is established outside the EU,
the EU authorized representative is:
Springer Nature Customer Service Center GmbH
Europaplatz 3, 69115 Heidelberg, Germany

Printed by Libri Plureos GmbH
in Hamburg, Germany